Intelligent Data Analysis for Biomedical Applications

Intelligent Data Analysis for Biomedical Applications
Challenges and Solutions

Dr. D. Jude Hemanth
Department of ECE, Karunya University, India

Dr. Deepak Gupta
Department of Computer Science and Engineering Maharaja Agrasen Institute of Technology, Guru Gobind Singh Indraprastha University, India

Prof. Valentina Emilia Balas
'Aurel Vlaicu' University of Arad, Romania

Series Editor Fatos Xhafa
Universitat Politècnica de Catalunya, Spain

ACADEMIC PRESS
An imprint of Elsevier

Academic Press is an imprint of Elsevier
125 London Wall, London EC2Y 5AS, United Kingdom
525 B Street, Suite 1650, San Diego, CA 92101, United States
50 Hampshire Street, 5th Floor, Cambridge, MA 02139, United States
The Boulevard, Langford Lane, Kidlington, Oxford OX5 1GB, United Kingdom

Notices
Knowledge and best practice in this field are constantly changing. As new research and experience
broaden our understanding, changes in research methods, professional practices, or medical treatment
may become necessary.

Practitioners and researchers must always rely on their own experience and knowledge in evaluating and
using any information, methods, compounds, or experiments described herein. In using such information
or methods they should be mindful of their own safety and the safety of others, including parties for
whom they have a professional responsibility.

To the fullest extent of the law, neither the Publisher nor the authors, contributors, or editors, assume any
liability for any injury and/or damage to persons or property as a matter of products liability, negligence
or otherwise, or from any use or operation of any methods, products, instructions, or ideas contained in
the material herein.

British Library Cataloguing-in-Publication Data
A catalogue record for this book is available from the British Library

Library of Congress Cataloging-in-Publication Data
A catalog record for this book is available from the Library of Congress

ISBN: 978-0-12-815553-0

For Information on all Academic Press publications
visit our website at https://www.elsevier.com/books-and-journals

Working together
to grow libraries in
developing countries

www.elsevier.com • www.bookaid.org

Publisher: Mara Conner
Acquisition Editor: Sonnini R. Yura
Editorial Project Manager: Naomi Robertson
Production Project Manager: Surya Narayanan Jayachandran
Cover Designer: Victoria Pearson

Typeset by MPS Limited, Chennai, India

Contents

List of Contributors

Mohammad Wajih Alam Department of Electrical and Computer Engineering, University of Saskatchewan, Saskatoon, SK, Canada

Sarah A. Alsuhaibani College of Computer Science and Information Technology, Imam Abdulrahman Bin Faisal University, Dammam, Saudi Arabia

Theodoros Anagnostopoulos Department of Infocommunication Technologies, ITMO University, St. Petersburg, Russia

J. Anitha Department of Electronics and Communication, Karunya University, Coimbatore, India

M. Bhuyan Department of Electronics and Communication Engineering, Tezpur University, Tezpur, India

J. Chandra Christ University, Bangalore, India

Bernardo De Elía Escuela Superior Técnica del Ejército, Buenos Aires, Argentina

Daniela López De Luise CIS Labs, Buenos Aires, Argentina

Wan Mahani Hafizah Department of Electronic Engineering, Faculty of Electrical and Electronics Engineering, Universiti Tun Hussein Onn Malaysia, Pt. Raja, Malaysia

A. Hazarika Department of Electronics and Communication Engineering, Tezpur University, Tezpur, India

Mohammed Imran College of Computer Science and Information Technology, Imam Abdulrahman Bin Faisal University, Dammam, Saudi Arabia

Muhammad Mahadi Abdul Jamil Department of Electronic Engineering, Faculty of Electrical and Electronics Engineering, Universiti Tun Hussein Onn Malaysia, Pt. Raja, Malaysia

Vijay Jeyakumar Department of Biomedical Engineering, SSN College of Engineering, Chennai, India

S.R. Jino Ramson Department of Electronics and Communication Engineering, Vignan's Foundation for Science, Technology and Research, Guntur, India

Mohd Farid Johan Department of Hematology, School of Medical Sciences, Universiti Sains Malaysia, Kubang Kerian, Malaysia

Bommannaraja Kanagaraj Department of Electronics and Communication Engineering, KPR Institute of Engineering and Technology, Coimbatore, India

G.R. Karpagam Department of CSE, PSG College of Technology, Coimbatore, India

Alimul H. Khan Department of Electrical and Computer Engineering, University of Saskatchewan, Saskatoon, SK, Canada

K. Lova Raju Department of Electronics and Communication Engineering, Vignan's Foundation for Science, Technology and Research, Guntur, India

Marcos Maciel CAETI, Buenos Aires, Argentina

Juan Pablo Menditto Escuela Superior Técnica del Ejército, Buenos Aires, Argentina

R. Merjulah Christ University, Bangalore, India

Manasa Nadipally JNTU Hyderabad, Telangana, India

Laghouiter Oussama Department of Electronic Engineering, Faculty of Electrical and Electronics Engineering, Universiti Tun Hussein Onn Malaysia, Pt. Raja, Malaysia

Eben Sophia Paul Department of ECE, Karpagam University, Coimbatore, India

Lucas Rancez CIS Labs, Buenos Aires, Argentina

N.J. Sairamya Karunya Institute of Technology and Sciences, Coimbatore, India

Md Hanif Ali Sohag Department of Electrical and Computer Engineering, University of Saskatchewan, Saskatoon, SK, Canada

M.S.P. Subathra Karunya Institute of Technology and Sciences, Coimbatore, India

Tanin Sultana Department of Electrical and Computer Engineering, University of Saskatchewan, Saskatoon, SK, Canada

L. Susmitha Karunya Institute of Technology and Sciences, Coimbatore, India

S. Thomas George Karunya Institute of Technology and Sciences, Coimbatore, India

B. Vinoth Kumar Department of IT, PSG College of Technology, Coimbatore, India

S. Vishnu Department of Electronics and Communication Engineering, Vignan's Foundation for Science, Technology and Research, Guntur, India

Mohd Helmy Abd Wahab Department of Computer Engineering, Faculty of Electrical and Electronics Engineering, Universiti Tun Hussein Onn Malaysia, Pt. Raja, Malaysia

Khan A. Wahid Department of Electrical and Computer Engineering, University of Saskatchewan, Saskatoon, SK, Canada

Yanjun Zhao Department of CS, Troy University, Troy, AL, United States

Chapter 1

IoT-Based Intelligent Capsule Endoscopy System: A Technical Review

Mohammad Wajih Alam, Md Hanif Ali Sohag, Alimul H. Khan, Tanin Sultana and Khan A. Wahid
Department of Electrical and Computer Engineering, University of Saskatchewan, Saskatoon, SK, Canada

Chapter Outline

1.1 INTRODUCTION

Wired endoscopy systems have been widely used to diagnose and monitor abnormalities in the gastrointestinal (GI) tract, such as obscure GI bleeding, Crohn's disease, cancer, and celiac disease [1,2]. Although effective and reliable, traditional endoscopy may cause discomfort and introduce complications in patients as this process requires a long and flexible tube to be pushed into the GI tract [3]. In addition, it is difficult to monitor certain

Intelligent Data Analysis for Biomedical Applications. DOI: https://doi.org/10.1016/B978-0-12-815553-0.00001-X

areas of the GI tract, such as the largest part of the small intestine [4]. Also, these endoscopes need trained professionals to operate them, which further requires a long time [5]. As a result of technological progress and successful clinical demonstrations, completely noninvasive endoscopic systems requiring no sedation have become a reality and are now commercially available for the diagnosis of various GI disorders.

A typical wireless capsule endoscopy (WCE) system consists of a pill-shaped electronic capsule, sensor belt, data recorder, and a workstation computer with image-processing software, as illustrated in Fig. 1.1. This electronic capsule is integrated with an image sensor, illumination optics, processing unit, communication modules, and batteries. The main features of various commercially available capsules are summarized in Table 1.1. It is clear from the table that image, sensor-based capsules are more abundant in the market. Other capsules utilize different sensors, such as a temperature sensor, pH sensor, or pressure sensor, to measure different physiological parameters. Although, the capsule endoscopy system has gained much popularity and shown its effectiveness, there are still significant limitations as evident from Table 1.1. The major limitations are lower battery-life, suboptimal image quality, lack of localization, and active locomotion control.

Integration of different sensors with the emerging Internet of Things (IoT) technology may enhance the existing functionality to a greater extent. Fig. 1.2 categorizes the additional features that can be introduced in the WCE system with the help of IoT based WCE system in the future. Each of these functionalities is discussed in detail in the following sections.

The remainder of this chapter is organized into four sections. Section 1.2 presents the data acquisition system, while Sections 1.3, 1.4, and 1.5 discusses the processing unit, data management, and proposed IoT-based WCE system, respectively.

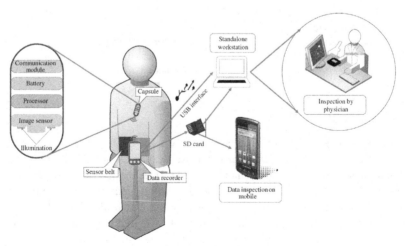

FIGURE 1.1 A typical block diagram of a wireless capsule endoscopy.

TABLE 1.1 Main Features of Commercially Available WCE Capsules

WCE Capsules Based on Image Sensor

Device	Company	Target Region	Size (mm)	Mass (g)	IL (LEDs)	BL (h)	FoV (°)	FR	Image Sensor	Res.	TM	RT
PillCam SB	Given imaging/ medtronic	Small bowel	26 × 11	< 4	6	8	140	2	1 CMOS	256 × 256	RF	Yes
PillCam SB2		Small bowel	26 × 11	2.9	4	8	156	2 or 4	1 CMOS	256 × 256	RF	Yes
PillCam SB3		Small bowel	26 × 11	3	4	12	156	2 or 2–6	1 CMOS	256 × 256	RF	Yes
PillCam ESO		Esophagus	26 × 11	2.9	6	0.5	172	35	2 CMOS	256 × 256	RF	Yes
PillCam ESO2		Esophagus	26 × 11	2.9	2 × 4	0.5	169	18	2 CMOS	256 × 256	RF	Yes
PillCam Colon		Colon	31 × 11	2.9	2 × 6	10	172	4	2 CMOS	256 × 256	RF	Yes
PillCam Colon2		Colon	31 × 11	2.9 ± 1	2 × 4	10	360	4–35	2 CMOS	256 × 256	RF	Yes
Miro Cam	Intromedic	Small bowel	24.5 × 10.8	3.25	6	12	170	3	1 CMOS	320 × 320	HBC	Yes
Endo Capsule	Olympus	Small bowel	26 × 11	3.3	6	12	160	2	1 CCD	1920 × 1080	RF	Yes
OMOM	JinShan	Small bowel	27.9 × 13	≤ 6	4	8 ± 1	140 ± 10	0.5 or 2	1 CMOS	640 × 480	RF	Yes
Capso Cam	Capso Vision	Small bowel	31 × 11	4	16	~15	360	3 or 5 p.c	4 CMOS	1920 × 1080	USB	No

WCE Capsule Based on Other Sensors

Device	Company	Target Region	Size (mm)	Mass (g)	BL (h)	Sensor(s)	Measuring Parameter	TM
SmartPill	Given Imaging/ Medtronic	GI tract	11.7 × 26.8	4.5	120	Pressure (0–350 mmHg), pH (0.05–9) and temperature (25–49°C)	Pressure, pH and temperature	RF
CorTemp	HQ Inc.	GI tract	10.37 × 22.32	2.75	168–240	Temperature sensor (30–45°C)	Temperature	RF
VitalSense	Philips	GI tract	8.7 × 23	1.6	240	Jonah™ core temperature sensors (25°C–60°C)	Core body temperature, dermal temperature, heart rate, respiration rate	RF
e-Celsius	BMedical Pty Ltd	GI tract	8.9 × 17.7	1.7	480	Temperature sensor (25°C–45°C)	Core body temperature	RF
Check-Cap	Check Cap Ltd	Colo-rectal	34 × 11.5	–	–	X-ray based 3D imaging	3-Dimensional lining of colon	RF

FoV: field of view; FR: frame rate; Res.: resolution; TM: transmission module; RT: real time; p.c: per camera; RF: radiofrequency; HBC: human body communication; IL: illumination; BL: battery life.

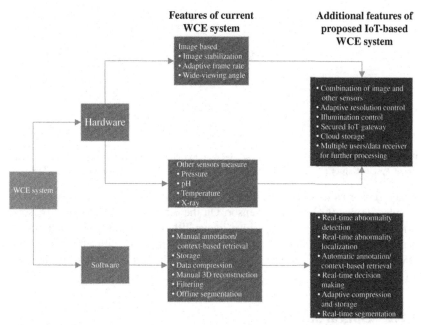

Features of current WCE system

Additional features of proposed IoT-based WCE system

WCE system

Hardware

Image based
• Image stabilization
• Adaptive frame rate
• Wide-viewing angle

Other sensors measure
• Pressure
• pH
• Temperature
• X-ray

• Combination of image and other sensors
• Adaptive resolution control
• Illumination control
• Secured IoT gateway
• Cloud storage
• Multiple users/data receiver for further processing

Software

• Manual annotation/ context-based retrieval
• Storage
• Data compression
• Manual 3D reconstruction
• Filtering
• Offline segmentation

• Real-time abnormality detection
• Real-time abnormality localization
• Automatic annotation/ context-based retrieval
• Real-time decision making
• Adaptive compression and storage
• Real-time segmentation

FIGURE 1.2   Comparison between current and the proposed IoT-based WCE system.

1.2   DATA ACQUISITION

Data acquisition in capsule endoscopy is performed using sensors. Since, the system is image-based, an image sensor is the main component. However, other sensors such as pH, temperature, pressure, and motion sensors have also been used. Usually, the outputs from sensors are analyzed, and actions are taken based on the analysis [6].

Determination of the GI motility, stricture, and produced gas in different regions of the GI tract could be used to diagnose various abnormalities. Moreover, ultrasonic reflection from the GI wall can provide detailed information about each layer, such as mucosa and submucosa. Besides these, the localization of abnormalities is an important issue which is still being researched. Various types of sensors could be utilized to address these issues which are discussed further in the following sub-sections. Fig. 1.3 shows some capsules that are available commercially.

1.2.1   Image Sensor

Conventional endoscopy systems use an image sensor to capture images of the GI tract. The image sensor (camera) along with illumination, processor, and wireless communication is miniaturized into a capsule. The most popular image sensor is the complementary metal oxide semiconductor (CMOS)

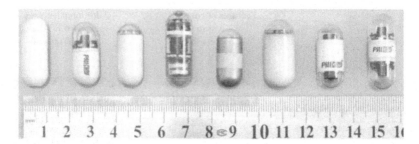

FIGURE 1.3 Commercial capsule endoscopes. Left to right: Agile patency capsule, PillCam SB2, EndoCapsule, CapsoCam, MiroCam, OMOM capsule, PillCam ESO, and PillCam COLON2 [81].

image sensor as it is cheaper, smaller, and consumes less power than charge-coupled device (CCD) image sensor. On the other hand, CCD image sensors offer a high quality and low-noise image [9]. Among the few small-bowel capsule endoscope models available on the market, PillCam, MiroCam, OMOM, and CapsoCam use CMOS imagers whereas the EndoCapsule uses a CCD imager [10].

1.2.2 Optical Sensor

The optical sensor measures different optical properties such as reflection, transmission, scattering, and absorption. The HemoPill includes an optical sensor that measures the intensity of transmitted light through a sample. The transmitted intensities of red and violet light are compared to detect acute bleeding [11,12]. The RGB color sensor can also be used to detect bleeding [13,14].

1.2.3 Pressure, Temperature, and pH-Monitoring Sensor

These sensors are used for motility monitoring. For instance, a pH profile can be useful to measure the transit time of the capsule. SmartPill (Fig. 1.3B) comprises all three sensors [15,16]. The temperature sensor is used in a few commercial capsules, such as, CorTemp [17], VitalSense [18], and e-celsius [19,20].

1.2.4 Other Ingestible Sensors

The odometer is a sensor that can measure the traveled distance of the capsule. OdoCapsule (Fig. 1.3C) uses this sensor to improve lesion localization [7,21]. Magnetic sensors could also be used to localize the capsule in real-time [22]. Ultrasound imaging in capsule endoscopy can be used to determine GI diseases. An ultrasound sensor emits ultrasound and reads the reflected sound from the GI wall. Mucosa, submucosa, and other layers of the GI wall have their own ultrasound profiles at normal states. However, abnormal profiles in any of the layers provide information about different types of diseases, such as tumors and cancer [23−26]. A gas sensor is used to get information about the chemical composition of the gut. A capsule has been developed by RMIT University, Australia that can sense oxygen, hydrogen, and carbon dioxide in the gut (Fig. 1.3D). This capsule could be used to evaluate the intestinal transit time, fermentative patterns, food modulation, drug disposition, intestinal physiology, and the effects of diet and medical supplements [8,27]. Radiofrequency identification (RFID) sends out an electronic signal which is detected by an external RFID scanner. The capsule can be localized using the RFID sensor. This sensor is used in the capsule to detect any stricture inside the GI tract to minimize the risk of occlusion. The patency capsule (Fig. 1.3A) uses a RFID sensor [28].

1.3 ON-CHIP DATA-PROCESSING UNIT

The acquired data from the sensors are processed by an on-chip data processor. A typical processing unit of a WCE capsule based on an image sensor is illustrated in Fig. 1.4. The required data is compressed and processed within the processing unit and later sent to the data recorder through the RF channel. The image sensor data can be compressed using different compression algorithms.

1.3.1 Image Compression

For detailed diagnosis and examination of the digestive diseases inside the GI tract, higher image resolution and frame rate are desired [34,35]. This will eventually increase the bit rate and power consumption of the RF transmitter. Therefore, efficient image compression techniques are required to compress the data, while maintaining an acceptable reconstruction quality of the source image. In a recent review [36], Alam et al. summarized the existing compression algorithms for WCE and also suggested some new techniques that can be used for future applications.

Turcza and Duplaga, in 2007 [37], proposed an image compression technique based on an integer version of discrete cosine transform (DCT) and the Huffman entropy encoder, which can produce both lossless and high-quality lossy compression. It is also suitable for simpler hardware implementation with limited power consumption. In 2006 [38], Lin et al. developed a

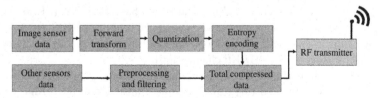

FIGURE 1.4 A typical processing unit.

low-powered, video compressor for GICam which consumes 14.92 mW and reduced the video size by a minimum of 75% and achieved RF transmission rate of 2 Mbps using the simplified traditional video compression algorithms with Lempel−Ziv (LZ) entropy coding [39] and a scalable compression architecture. Khan and Wahid [40] proposed an image compression algorithm based on a static prediction scheme and combination of golomb-rice and unary encoding with a compression ratio close to 73%. It requires lower computational complexity and 18 μW of power while working at 2 fps.

It can be seen that there exists a number of data compression techniques. The image compression algorithm in the capsule must consume a minimal amount of power to facilitate imaging of the entire GI tract. For this reason, a lossy image compression algorithm is usually selected to maintain a trade-off between the image quality and the frame rate, while minimizing the required physical size of the microchip and its power consumption.

1.3.2 Application Specific Integrated Circuit Design

For the proper implementation of image compression algorithms, as well as efficient processing of the various sensor data inside the capsule, the Application Specific Integrated Circuit (ASIC) chip plays a vital role. It also controls the transmission of the image sensor data using an RF channel and can take command of the RF channel to take action accordingly using a bidirectional communication method [41]. Predominant enhancement in the ASIC configuration can reduce the power utilization of the system and increase the frame rate of the image sensor. This chip was initially designed by Zarlink Semiconductor, Inc.

A low-power, near-field transmitter for capsule endoscopy was developed by Thone et al. [42]. They designed a low-power, near-field transmitter for the WCE system at 144 MHz carrier frequency to minimize the attenuation loss and to reduce power consumption to only 2 mW for transmitting the compressed image data at the rate of 2 Mbps. The antenna system size can also be reduced to a greater extent by using a high-frequency carrier. Goa et al. reduced the antenna size further with an ultra-wideband (UWB) (3−5 GHz), low-power telemetry transceiver system with a 0.18 μm CMOS process of 4×3 mm^2 outer dimensions and transmitted the image data at a rate of 10 Mbps with a 1.8 V power supply [43,44]. Diao et al. further improved the design by increasing the data rate to 15 Mbps at 900 MHz using an industrial, scientific and medical (ISM) band [45]. Later, Kim et al. designed a high-efficiency transceiver system of 0.13 μm CMOS process in 1.0 mm^2 silicon area with a simple on−off key transmitter having a data rate of 20 Mbps using a 500 MHz RF channel [46].

Integrated chips handle the image compression algorithms to increase the frame rate using the existing low data-rate transmission line. Khan and Wahid proposed a low-power, low-complexity compressor for capsule

FIGURE 1.5   Prototype of an FPGA-based capsule with smart-device connectivity [47].

endoscopy [40] which can achieve a compression ratio of 80% using only 42 μW consumption power. Fig. 1.5 shows a prototype of an FPGA-based capsule with smart-device connectivity.

1.3.3   Radiofrequency Transmission

After processing and compressing the image sensor data, it is sent to the data logger using an RF transmitter. The receiver/recorder unit receives and records the images through an antenna array consisting of several leads that are connected by wires to the recording unit, which is worn in standard locations over the abdomen suitable for lead placement. The recording device to which the leads are attached is capable of recording images which are transmitted by the capsule and received by the antenna array. Considering the human body as a lossy dielectric media that absorbs the waves and attenuates the receiving signal, it presents a strong negative effect to the microwave propagation. Hence, the quality of the received images and the power consumption of the battery depend on the signal-transmission efficiency of the antenna [48].

The role of the embedded antenna is to transmit the detected signals from inside the body to the receiver outside the human body. The ideal antenna for WCE should be less sensitive to human tissue and must have enough bandwidth to transmit high-resolution images along with a high-data rate [48]. Spiral, double-arm spiral, conical helix, fat-arm spiral, square microstrip loop, etc., are the types of antennas which are generally used in commercial WCE as well as in research [48]. A spiral transmitting and receiving antenna is shown in Fig. 1.6.

The mode of data transmission currently uses ultra-high frequency (UHF) band radio telemetry (e.g., PillCam, EndoCapsule). The human body communications (HBC) used by MiroCam is another type of transmission mode that utilizes the capsule itself to generate an electrical field that uses human tissue as the conductor for data transmission [50]. Inductive, link-based designs typically use a frequency transmission of 20 MHz or lower [51].

Using low frequency can achieve high transmission efficiency through layers of human skin [50]. The 402 MHz (UHF) Medical Implant Communication Service is a global, license-free service that has a small

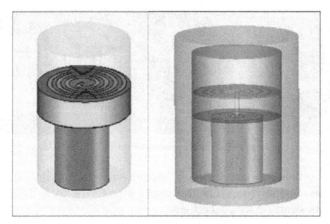

FIGURE 1.6 Spiral transmitting and receiving antenna [49].

bandwidth of 300 kHz, which is insufficient for video imaging-based WCE application as they require high-data rate and high resolution. The 2.45 GHz ISM band, on the other hand, offers a larger bandwidth [52] although this is still not enough due to impedance mismatch in a wide bandwidth. One of the possible frequency bands to provide high-resolution images from WCE is the use of UWB at the frequency of 3.1−10.6 GHz [53]. Hence, the selection of a proper operating frequency and transmission channel has received significant attention in current research. The different modulation techniques that have been used in RF telemetry are Frequency Shift Keying, Amplitude Modulation, On−Off keying, and Binary Phase Shift Keying, etc. [51] IEEE C95.1−2005 is the human exposure standard in RF radiation. It stands for Standard for Safety Levels with Respect to Human Exposure to Radiofrequency Electromagnetic Fields, 3 kHz to 300 GHz.

1.3.4 Power Management

The power management unit mainly consists of processing logic and power sources, such as batteries. Frame rate modulation techniques can also be used to control the frame rate within the desired area to ensure the capsule's longer operating time [54]. Smart Motion Sense Technology can also be helpful to activate or deactivate the capsule camera by sensing the motion within the GI tract [55]. Synchronous switching of the light-emitting diode (LED) and the CMOS sensors can also minimize power consumption. An intelligent energy efficient system was proposed by Liu et al., which can shut down and wake up necessary components of the capsule when needed [56].

Moreover, there will be many other sensors along with the image sensor to provide additional functionalities, like bleeding detection, active system control over the capsule's motion, gas sensing, active locomotion, and localization etc., in the more-advanced IoT-based WCE system. However, these will require much more power which may not be fulfilled by a conventional power source that can merely deliver around 25 mW [57,58]. Harvesting energy from different ambient sources, such as radiofrequency, light, vibrations, or thermoelectricity, can provide electrical energy in the range of μW, which is still very low for a WCE system [59]. Therefore, wireless power transmission (WPT) may be the best alternative to overcome the power limitations of an onboard battery system. In case of WPT, a transmitting coil is kept at a certain distance and appropriate location outside the human body and the receiving coil is placed within the WCE. The WPT was first tested by RF System Lab on their capsules (Norika and Sayaka) which did not need any battery power [60]. One group has proposed primary magnetic coils in a power-generating device outside the body to send power to a capsule within the body to save space [61]. Lenaerts and Puers introduced an inductive power link which was capable of transmitting 150 mW [62]. Recent research on WPT has improved significantly and can easily transmit 500 mW of power [63]. The WPT also offers the flexibility of adjusting the required amount of power for the capsule. Hence, it could be said that the WPT would be the best option to provide relatively higher levels of power to an active WCE system with multiple sensors.

1.4 DATA MANAGEMENT OF WIRELESS CAPSULE ENDOSCOPY SYSTEMS

The WCE system still relies greatly on workstation software. A typical endoscopy capsule, when administered into the human body, works for about 8−10 h and generates around 57,600 frames [64]. A doctor/physician is still required to sit in front of the computer to analyze each frame which is tiresome and inefficient. Although some functionalities exist that help the physician for automating the detection of abnormalities, the existing features are still inefficient and unreliable. During postprocessing, the data recorded on the data logger is retrieved at a workstation where software is installed to analyze the recorded data. This software is provided by the corresponding vendor. Different vendors of WCE have developed various software with different functionalities which are maintained and updated regularly, for example, RAPID Reader software is developed and provided by PillCam.

This software includes some advanced functionalities that help physicians with automating detection of bleeding. In particular, the suspected blood indicator helps in distinguishing red pixels of the captured images by image sensor automatically. The sensitivity for identification of active bleeding is reported to be in the range of 58.3%−93% [65,66]. These results reveal that this software cannot replace physicians; however, these additional features can definitely help physicians to detect bleeding inside the GI tract and save time.

Many researchers have been working on automating the detection of abnormalities inside the GI tract during postprocessing. Li et al. [67] proposed the novel idea of chrominance as a part of the color texture feature, which utilizes Tchebichef polynomials and an illumination invariant of Hue/Saturation/Intensity color space which is used to differentiate between normal and bleeding regions. Similarly, Pan et al. [68] proposed a computer-aided system based on a neural network where color texture features of bleeding regions are extracted and a probabilistic classifier is built to detect bleeding.

1.5 IOT-BASED WIRELESS CAPSULE ENDOSCOPY SYSTEM

With a view to provide better patient care, reduced health service cost, and improved treatment outcomes, IoT is constantly offering new opportunities to enhance the integrated health-care field either by introducing new scopes of health-care facilities or by improving existing systems of health care [71]. Current common hospital-centered health care will be transformed from, first to hospital-home-balanced care by 2020, and ultimately, to home-centered care by 2030 [72]. The current WCE system cannot yet offer real-time detection as the diagnosis is done offline. In addition, a self data-analyzing intelligent system could be helpful, which can select useful data for transmission and save transmission cost, bandwidth, and energy. Real-time viewing may also assist in capsule manipulation and drug delivery in the future [73]. To develop IoT-based systems, characteristics like intelligence, heterogeneous network connectivity, real-time sensing capability, and security may be incorporated into the system.

1.5.1 Intelligence in the System

Intelligent decision-making capability is a built-in key utility factor of any IoT-associated system [74]. Current research shows that intelligent features, such as the adaptive frame rate and the wide viewing angle, can be implemented in the capsule when used in real-time [75]. An IoT-based system can make instant decisions by making a choice between useful and redundant frames. Besides, it can efficiently detect selected anomalies inside the GI tract based on predefined algorithms.

1.5.2 Real-Time Sensing

The instant decision-making capability of the system requires real-time sensing abilities that an IoT-based system can offer [63,74]. Real-time monitoring allows physicians to take quick measures when required. Besides, with the advent of real-time viewing and external manipulation, the notion of targeted drug delivery (targeted therapeutics) becomes feasible [76]. Capsule propulsion needs to be followed by real-time viewing during the first hour to make sure the capsule passes the stomach [55].

1.5.3   Internet of Things Protocol

Compared to the traditional networking protocols, the dedicated IoT protocols are lightweight in nature, consume less bandwidth, less power (Message Queuing Telemetry Transport (MQTT) and Constrained Application Protocol (CoAP) are familiar as constrained application protocols). Hence, IoT communication protocols are best-suited for low power, constraint environment devices like WCE to transfer various sensor data in real time. Besides this, IoT protocols offer bidirectional control over sensors [77]. For example, the resolution or frame rate may be adjusted for diagnostically important frames.

1.5.4   Connectivity

IoT devices may support several interoperable communication protocols (for example CoAP can interoperate with Hyper Text Transfer Protocol (HTTP) [77]) and can communicate with other devices and infrastructure [74]. It also provides less, or no, latency in communication. In fact, the smart IoT gateways can act as a hub between wireless body/personal/local area networks (WBAN/WPAN/WLAN) and a remote health-care center [78]. This feature can successfully integrate remote monitoring of the endoscopy capsule so that the data acquired can be accessed by qualified physicians from anywhere in the world.

1.5.5   Security

Ensuring the confidentiality and privacy of medical images is becoming a challenging problem as these images contain sensitive data with distinguishing visual representations of the interior of a human body [71]. Current research shows that it is possible that this image data can be transferred through Wi-Fi remotely using a smartphone as the workstation [79]. Therefore, it is necessary to send the frames securely. An IoT-based system provides functions such as authentication, authorization, privacy, message integrity, content integrity, and data security [74]. For example, MQTT uses Transport Layer Security/Secure Sockets Layer (TLS/SSL) and CoAP uses Data-gram Transport Layer Security (DTLS) for security [77].

1.5.6   Improved Outcomes of Treatment

The automatic connectivity through cloud computing or another virtual infrastructure gives physicians the ability to access real-time information that enables them to make informed decisions. This ensures improved health care and treatment outcomes through reduced human contact with patients [80].

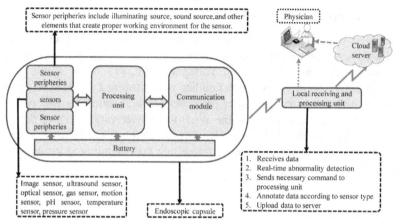

FIGURE 1.7   Proposed IoT-based future WCE system.

The proposed IoT-based WCE system is illustrated in Fig. 1.7. The system connects the body sensor network that consists of multiple sensors as described in Section 1.2. The sensor peripheries will help to create a proper working environment for the corresponding sensor, for example, an image sensor requires an illumination source (LED), and an ultrasound sensor may require a sound source, etc. The sensor data will be processed with the help of the processing unit which will be transmitted with the help of the communication module to the local processing and receiving unit that consists of the IoT gateway and local processing unit outside the body. The communication module can be a bidirectional RF transceiver. The IoT local processing unit can analyze and annotate different types of sensor data according to the abnormality. It can also send commands to the processing unit inside the capsule, for example, when an abnormality is detected in an image, it can send a command to increase the frame rate, resolution, or illumination, etc. By extracting the information and sorting them, the IoT gateway can further proceed these data toward the remote area network using IoT protocols like MQTT or CoAP, etc., with the built-in security (like DTLS or TLS). The receiver of these data can be a dedicated WCE cloud server from which an authorized physician can access the data in real-time from anywhere by using either PC, phone, or website. The WCE cloud server will store all data of the corresponding patient for further use. Finally, the intelligence of IoT can efficiently accelerate detection and help physicians to receive the necessary information to diagnose the actual disease type quickly.

1.6   FUTURE CHALLENGES

While some of these sensors are used in commercial pills, many new sensors can be integrated to improve the tool's effectiveness and diagnostic accuracy.

Localization is an important issue for the treatment of GI diseases. The pH and pressure profile of the human GI tract could be used to localize the capsule based on different GI profiles, such as when it leaves the stomach, duodenum, or jejunum. The motion sensor and RFID can also provide sufficient information about localization [29,30]. Optical sensors are able to determine bleeding and can be used to detect other abnormalities of the GI tract, such as the lesions, tumors, and cancerous cells [31,32]. Other imaging techniques, such as the ultrasound imaging, fluorescence imaging [33], and X-ray imaging could provide additional, in-depth information of the gut. Moreover, the possibility of missing the abnormalities by traditional technologies will be reduced by using multiple sensors. By using these sensors, it is also possible to prolong battery life to several days [27]. Future WCE software should have an additional feature to distinguish between normal and abnormal profiles.

Data management is necessary in order to efficiently gain the information that the user is looking for, either in a single video of a trial or in an archive. The software should be able to respond to a specific index-based query which lists the desired information. This will help with efficient content exploration.

The current scenario to retrieve information from a large video is to manually annotate each frame. However, manual annotation is not always a feasible option when there is a large video, especially if the physician has a limited time. Hence, it will be interesting to utilize the technique used by doctors to extract the important information from a limited number of frames and apply the same technique for the remaining video archives. The content-based, image-retrieval system will help in distinguishing between images of interest to those of no interest.

In terms of compression and storage, only the frames of interest can be compressed and stored. Munzer et al. [69] showed that the circular segment of the endoscopic images (which is usually dark, blurry, or noisy) can be discarded for improving encoding efficiency and saving 20% storage space. Lossless compression can be incorporated for the region of interest while lossy compression can be used for other regions [36]. Besides these, storing the videos in HD quality is not always required; in some cases a lower quality of images still provides enough clarity for subjective evaluation [70].

1.7 CONCLUSION

The objective of this chapter is to discuss the limitations of the current WCE system and suggest possible remedies. The WCE tool plays an important role in diagnosing GI tract disorders through a painless procedure. It is a safe procedure and widely used for small-bowel screening, where a wired endoscope cannot reach. However, the system lacks illumination and resolution control, localization of abnormality, and real-time decision-making, etc. Therefore, an improved WCE system is needed which must resolve the challenges of data acquisition, data storage and processing, and data analysis as major blocks. The data acquisition block consists of various sensors, such as image, gas, temperature, optical, ultrasound, pH, and motion sensors. These sensors aid in improving various functionalities of the current capsule endoscope, for instance by detecting and localizing abnormality inside the GI tract, real-time sensing, and increasing the quality of the images, etc.

The inclusion of an IoT platform will enable real-time diagnosis as well as adaptive illumination and resolution control. Analyzing data using the current capsule endoscopic system is another challenging task for a physician, as they are required to sit in front of a computer for 8−10 h to go through each frame. Although many image segmentations or auto-detection algorithms are already available, these features are offline. Thus a system which can automatically annotate the frames is needed which will help save time for physicians. It will also aid in saving storage space during postprocessing, as discussed in Section 1.4, making the system more efficient.

Considering the current limitations, it is essential that the IoT-based capsule endoscopy system is developed to improve the diagnostic efficiency of GI abnormalities. Improved imaging, adaptive resolution control, and an improved illumination source, as well as the development of novel tools for automating detection and the localization of abnormality, smart data management system utilizing IoT protocols are some of the important goals that need to be addressed in the development of future capsule endoscopy systems.

REFERENCES

[1] G. Ciuti, A. Menciassi, P. Dario, Capsule endoscopy: from current achievements to open challenges, IEEE Rev. Biomed. Eng. 4 (2011) 59−72.

[2] M.R. Basar, F. Malek, K.M. Juni, M.S. Idris, M.I.M. Saleh, Ingestible wireless capsule technology: a review of development and future indication, Int. J. Antennas Propag. 2012 (807165) (2012) 1−14.

[3] G. Iddan, G. Meron, A. Glukhovsky, P. Swain, Wireless capsule endoscopy, Nature 405 (6785) (2000) 417.

[4] A. Loeve, P. Breedveld, J. Dankelman, Scopes too flexible and too stiff, IEEE Pulse 1 (6) (2010) 2154−2287.

[5] A. Menciassi, M. Quirini, P. Dario, Microrobotics for future gastrointestinal endoscopy, Minim. Invasive Therapy Allied Technol. 16 (2) (2007) 91−100.

[6] I.R. Sinclair, Sensors and Transducers, third ed., Newness, Oxford, 2001.

[7] A. Karargyris, A. Koulaouzidis, OdoCapsule: next generation wireless capsule endoscopy with accurate localization and video stabilization, IEEE Trans. Biomed. Eng. 62 (1) (2014) 352−360.

[8] K. Zadeh, et al., A human pilot trial of ingestible electronic capsules capable of sensing different gases in the gut, Nat. Electron 1 (1) (2018) 79−87.

[9] M. Bigas, E. Cabruja, J. Forest, J. Salvi, Review of CMOS image sensors, Microelectron. J. 37 (5) (2006) 433−451.

[10] L.J. Sliker, G. Ciuti, Flexible and capsule endoscopy for screening, diagnosis and treatment, Expert Rev. Med. Dev. 11 (6) (2014) 649−666.

[11]   S. Schostek, M.O. Schurr, The HemoCop telemetric sensor system-technology and result of in-vivo assessment, Stud. Health Technol. Inform. 177 (2012) 97–100.

[12]   S. Schostek, et al., Telemetric real-time sensor for the detection of acute upper gastrointestinal bleeding, Biosens. Bioelectron. 78 (2016) 524–529.

[13]   P. Qiao, et al., A smart capsule system for automated detection of intestinal bleeding using HSL color recognition, PLoS ONE 11 (11) (2016) 1–14.

[14]   H. Liu, et al., An intelligent electronic capsule system for automated detection of gastrointestinal bleeding, J. Zhejiang Univ. Sci. B 11 (12) (2010) 937–943.

[15]   H.O. Diaz Tartera, et al., Validation of SmartPill® wireless motility capsule for gastrointestinal transit time: intra-subject variability, software accuracy and comparison with video capsule endoscopy, Neurogastroenterol. Motil. 29 (10) (2017) 1–9.

[16]   A.M. Stokes, et al., Evaluation of a wireless ambulatory capsule (SmartPill) to measure gastrointestinal tract pH, luminal pressure and temperature, and transit time in ponies, Equine Vet. J. 44 (4) (2012) 482–486.

[17]   CorTemp Sensor-HQInc. Available: http://www.hqinc.net/cortemp-sensor-2/ (accessed 18.02.18).

[18]   VitalSense Wireless Vital Signs Monitoring from Mini Mitter Company, Bend Oregon, Internet: http://www.temperatures.com/mmvs.html (accessed 18.02.18).

[19]   e-Celsius Medical-Bodycap, Internet: http://www.bodycapmedical.com/en/product/ecelsius (accessed 18.02.18).

[20]   G. Roussey, et al., A comparative study of two ambulatory core temperature assessment methods, IRBM 39 (2) (2018) 143–150.

[21]   A. Karargyris, A. Koulaouzidis, Capsule-odometer: a concept to improve accurate lesion localisation, World J. Gastroenterol. 19 (35) (2013) 5943–5946.

[22]   D.M. Pham, S.M. Aziz, A real-time localization system for an endoscopic capsule using magnetic sensors, Sensors (Basel) 14 (11) (2014) 20910–20929.

[23]   F. Memon, et al., Capsule Ultrasound Device, 2015 IEEE International Ultrasonics Symposium (IUS), Taipei, 2015, pp. 1–4.

[24]   J.H. Lee, et al., Towards Wireless Capsule Endoscopic Ultrasound (WCEU), IEEE Int. Ultrason. Symp. IUS, 2014, pp. 734–737.

[25]   B.F. Cox, et al., Ultrasound capsule endoscopy: sounding out the future, Ann. Transl. Med. 5 (9) (2017), pp. 201–201.

[26]   S. Wei, et al., An imaging device towards ultrasonic capsule endoscopy, Shenzhen Daxue Xuebao (Ligong Ban)/J. Shenzhen Univ. Sci. Eng. 33 (3) (2016) 259–263.

[27]   K. Zadeh, et al., Ingestible sensors, ACS Sens. 2 (4) (2017) 468–483.

[28]   Á.C. Álvarez, et al., Patency and agile capsules, World J. Gastroenterol. 14 (34) (2008) 5269–5273.

[29]   J.-C. Saurin, N. Beneche, C. Chambon, M. Pioche, Challenges and future of wireless capsule endoscopy, Clin. Endosc. 49 (1) (2016) 26–29.

[30]   C. Rommele, J. Brueckner, H. Messmann, S.K. Golder, Clinical experience with the PillCam patency capsule prior to video capsule endoscopy: a real-world experience, Gastroenterol. Res. Pract 2016 (2016). Article ID: 9657053, 6 pages.

[31]   M.A. Al-Rawhani, J. Beeley, D.R.S. Cumming, Wireless fluorescence capsule for endoscopy using single photon-based detection, Sci. Rep. 5 (2015) 18591.

[32]   P. Demosthenous, C. Pitris, J. Georgiou, Infrared fluorescence-based cancer screening capsule for the small intestine, IEEE Trans. Biomed. Circuits Syst. 10 (2) (2016) 467–476.

[33] M.M. Hasan, M.W. Alam, K.A. Wahid, S. Miah, K.E. Lukong, A low-cost digital micro-scope with real-time fluorescent imaging capability, PLoS ONE 11 (12) (2016) e0167863.

[34] M. Mylonaki, A. Fritscher-Ravens, P. Swain, Wireless capsule endoscopy: a comparison with push enteroscopy in patient with gastroscopy and colonoscopy negative gastrointesti-nal bleeding, Gut J. (2003) 1122−1126.

[35] D. Turgis, R. Puers, Image compression in video radio transmission for capsule endos-copy, Sens. Actuators A 123−124 (2005) 129−136.

[36] M.W. Alam, M.M. Hasan, S.K. Mohammed, F. Deeba, K.A. Wahid, Are current advances of compression algorithms for capsule endoscopy enough? A technical review, IEEE Rev. Biomed. Eng. 10 (2017).

[37] P. Turcza, M. Duplaga, Low-power image compression for wireless capsule endoscopy, Proc. of the IEEE Int. Workshop on Imaging Systems and Technology, pp. 1−4, 2007.

[38] M.C. Lin, L.R. Dung, P.K. Weng, et al., An ultra low power image compressor for cap-sule endoscope, Biomed. Eng. Online 5 (2006) 14.

[39] J. Ziv, A. Lempel, A. Universal, Algorithm for sequential data compression, IEEE Trans. Inf. Theory 23 (1977) 337−343.

[40] T.H. Khan, K.A. Wahid, Low power and low complexity compressor for video capsule endoscopy, IEEE Trans. Circuits Syst. Video Technol. 21 (2011) 1534−1546.

[41] A. Anpilogov, I. Bulychev, A. Tolstaya, Features of the data transmission in the wireless capsule endoscopic complex, Procedia Comput. Sci. 88 (2016) 312−317. Available from: https://doi.org/10.1016/j.procs.2016.07.441.

[42] J. Thone, S. Radiom, D. Turgis, R. Carta, G. Gielen, R. Puers, Design of a 2 Mbps FSK near-field transmitter for wireless capsule endoscopy, Sens. Actuators A 156 (1) (2009) 43−48.

[43] Y. Gao, Y. Zheng, S. Diao, et al., Low-power ultrawideband wireless telemetry trans-ceiver for medical sensor applications, IEEE Trans. Biomed. Eng. 58 (3) (2011) 768−772.

[44] Y. Gao, S. Diao, C.W. Ang, Y. Zheng, X. Yuan, Low power ultra-wideband wireless telemetry system for capsule endoscopy application, in: Proceedings of IEEE International Conference on Robotics, Automation and Mechatronics (RAM'10), pp. 96−99, sgp, 2010.

[45] S. Diao, Y. Gao, W. Toh, et al., A low-power, high data-rate CMOS ASK transmitter for wire-less capsule endoscopy, in: Defense Science Research Conference and Expo, pp. 1−4, 2011.

[46] K. Kim, S. Yun, S. Lee, S. Nam, Y. Yoon, C. Cheon, A design of a high-speed and high-efficiency capsule endoscopy system, IEEE Transact. Biomed. Eng. 59 (2012) 1005−1011.

[47] T.H. Khan, R. Shrestha, K.A. Wahid, P. Babyn, Design of a smart-device and FPGA based wireless capsule endoscopic system, Sens. Actuators A: Phys. 221 (2015) 77−87.

[48] Z. Wang, E.G. Lim, T. Tillo, F. Yu, Review of the wireless capsule transmitting and receiving antennas, wireless communications and networks, in: A. Eksim (Ed.), Recent Advances, InTech, 2012. Available from: https://www.intechopen.com/books/wireless-communications-and-networks-recent-advances/review-of-the-wireless-capsule-transmit-ting-and-receiving-antennas.

[49] J. Wang, Z. Wang, M. Leach, et al., RF characteristics of wireless capsule endoscopy in human body, J. Cent. South Univ. 23 (5) (2016) 1198−1207.

[50] A. Wang, S. Banerjee, B.A. Barth, Y.M. Bhat, S. Chauhan, K.T. Gottlieb, et al., Wireless capsule endoscopy, Gastrointest. Endosc. 78 (6) (2013) 805−815. ISSN 0016-5107.

[51] M.R. Yuce, T. Dissanayake, Easy-to-swallow wireless telemetry, IEEE Microwave Mag. 13 (6) (2012) 90−101.

[52]   R. Das, H. Yoo, A wideband circularly polarized conformal endoscopic antenna system for high-speed data transfer, IEEE Trans. Antennas Propag. 65 (6) (2017) 2816–2826.

[53]   K.Y. Yazdandoost, Antenna for wireless capsule endoscopy at ultra wideband frequency, in: 2016 IEEE 27th Annual International Symposium on Personal, Indoor, and Mobile Radio Communications (PIMRC), Valencia, 2016, pp. 1–5.

[54]   Z. Liao, C. Xu, Z.S. Li, Completion rate and diagnostic yield of small-bowel capsule endoscopy: 1 vs. 2 frames per second, Endoscopy 42 (2010) 360–364.

[55]   C. Van de Bruaene, D. De Looze, P. Hindryckx, Small bowel capsule endoscopy: where are we afer almost 15 years of use? World J. Gastrointest. Endosc. 7 (2015) 13–36.

[56]   G. Liu, G. Yan, B. Zhu, L. Lu, Design of a video capsule endoscopy system with low-power ASIC for monitoring gastrointestinal tract, Med. Biol. Eng. Comput. 54 (11) (2016) 1779–1791.

[57]   R. Carta, R. Puers, Wireless power and data transmission for robotic capsule endoscopes, in: Proceedings of the 2011 18th IEEE Symposium on Communications and Vehicular Technology in the Benelux (SCVT), Ghent, Belgium, 22–23 November 2011; pp. 1–6.

[58]   R. Carta, G. Tortora, J. Thoné, B. Lenaerts, P. Valdastri, A. Menciassi, et al., Wireless powering for a self-propelled and steerable endoscopic capsule for stomach inspection, Biosens. Bioelectron. 25 (2009) 845–851.

[59]   N.S. Hudak, G.G. Amatucci, Small-scale energy harvesting through thermoelectric, vibration, and radiofrequency power conversion, J. Appl. Phys. 103 (2008) 101301.

[60]   Sayaka capsule system, Available: http://www.rfsystemlab.com/en/sayaka/index.html (accessed 29.03.18).

[61]   W. Xin, G. Yan, W. Wang, Study of a wireless power transmission system for an active capsule endoscope, Int. J. Med. Robot. 6 (2010) 113–122.

[62]   B. Lenaerts, R. Puers, An inductive power link for a wireless endoscope, Biosens. Bioelectron. 22 (7) (2007) 1390–1395.

[63]   Z. Jia, G. Yan, H. Liu, Z. Wang, P. Jiang, Y. Shi, The optimization of wireless power transmission: design and realization, Int. J. Med. Robot. Comput. Assist. Surg. 8 (2012).

[64]   J.B. Montgomery, et al., Is there an application for wireless capsule endoscopy in horses, Can. Vet. J. 58 (12) (2017) 1321–1325.

[65]   S. Liangpunsakul, et al., Performance of given suspected blood indicator, Am. J. Gastroenterol. 98 (12) (2003) 2676–2678.

[66]   J.M. Buscaglia, et al., Performance characteristics of the suspected blood indicator feature in capsule endoscopy according to indication for study, Clin. Gastroenterol. Hepatol. 6 (3) (2008) 298–301.

[67]   B. Li, et al., Computer-aided detection of bleeding regions for capsule endoscopy images, IEEE Trans. Biomed. Eng. 56 (2009) 1032–1039.

[68]   G. Pan, et al., Bleeding detection in wireless capsule endoscopy based on probabilistic neural network, J. Med. Syst. 35 (2011) 1477–1484.

[69]   B. Munzer et al., Improving encoding efficiency of endoscopic videos by using circle detection based border overlays, in: IEEE International Conference on Multimedia and Expo Workshops (ICMEW), pp. 1–4, 2013.

[70]   B. Munzer et al., "Investigation of the impact of compression on the perceptual quality of laparoscopic videos", in: 2014 IEEE 27th International Symposium on Computer-based Medical Systems (CBMS), pp. 365–388, 2014.

[71] R. Hamza, K. Muhammad, L. Zhihan, F. Titouna, Secure video summarization framework for personalized wireless capsule endoscopy, Pervasive Mobile Comput. 41 (2017) 436–450.

[72] C.E. Koop, et al., Future delivery of health care: cybercare, IEEE Eng. Med. Biol. Mag. 27 (6) (2008) 29–38.

[73] P. Swain, The future of wireless capsule endoscopy, World J. Gastroenterol. 14 (26) (2008) 4142–4145.

[74] P.P. Ray, A survey on Internet of Things architectures, J. King Saud Univ. Comput. Inf. Sci. 30 (3) (2016) 291–319.

[75] S.N. Adler, Y.C. Metzger, PillCam COLON capsule endoscopy: recent advances and new insights, Therap. Adv. Gastroenterol. 4 (4) (2011) 265–268.

[76] M.F. Hale, R. Sidhu, M.E. McAlindon, Capsule endoscopy: current practice and future directions, World J. Gastroenterol. 20 (24) (2014) 7752–7759.

[77] N. Naik, Choice of effective messaging protocols for IoT systems: MQTT, CoAP, AMQP and HTTP, in: 2017 IEEE International Systems Engineering Symposium (ISSE), Vienna, 2017, pp. 1–7.

[78] Rahmani et al., Smart e-Health gateway: bringing intelligence to Internet-of-things based ubiquitous healthcare systems, in: 12th Annual IEEE Consumer Communications and Networking Conference (CCNC), Las Vegas, NV, pp. 826–834, 2015.

[79] R. Shrestha, T. Khan, K. Wahid, Towards real-time remote diagnostics of capsule endoscopic images using Wi-Fi, in: 2nd Middle East Conference on Biomedical Engineering, Doha, 2014, pp. 293–296.

[80] P.A. Laplante, N. Laplante, The Internet of things in healthcare: potential applications and challenges, IT Prof. IEEE Comput. Soc. 18 (3) (2016).

[81] M. Keuchel, F. Hagenmuller, H. Tajiri, "Video Capsule Endoscopy: A Reference Guide and Atlas", 1st ed. New York, NY, USA: Springer, 2014.

Chapter 2

Optimization of Methods for Image-Texture Segmentation Using Ant Colony Optimization

Manasa Nadipally
JNTU Hyderabad, Telangana, India

Chapter Outline

2.1 INTRODUCTION

Image segmentation is regarded as an integral component in digital image processing which is used for dividing the image into different segments and discrete regions. The outcome of image segmentation is a group of segments that jointly enclose the whole image or a collection of contours taken out

Intelligent Data Analysis for Biomedical Applications. DOI: https://doi.org/10.1016/B978-0-12-815553-0.00002-1

from the image. We can divide image segmentation into different methods. For instance, methods based on compression techniques propose that the best method of segmentation is the one which minimizes data's coding length and the general probable segmentations. Methods based on histograms are known to be extremely well-organized to evaluate additional segmentation schemes as they need only single exceed in the progression of the pixels [1]. In this scheme, all of the pixels of an image are taken into consideration to figure the histogram, and the valleys and peaks in the histogram are utilized for establishing the clusters in an image. Inside-image processing, edge detection is a robust field on its own. Region edges and boundaries are connected directly since there is often a quick modification in strength at the area of boundaries [2]. Another significant part of image processing is thresholding, which is used for conversion of a grayscale image into a binary image. For this purpose, the threshold value is chosen after selection of multiple-levels. We assume that an image is divided into the following two parts: foreground and background. The Hachemi Guerrout method exhibits its resistance and robustness to noise by employing a Hidden Markov Random Field-Particle Swarm Optimization (HMRF-PSO) technique over threshold-based schemes [3]. Liang and Leung have described a genetic algorithm with adaptive, exclusive, population tactics for function optimization in multimode [4]. Two other scientists, Wang and Huang [5], have described a thresholding method using a selection of an adaptive window for irregular lighting images. The Wang and Jiang methods explain color image segmentation whose basis is region-merging and homogram-thresholding [5]. A clustering technique based on a genetic algorithm has been proposed by the Maulik method [6]. Further, the Chang method has introduced a rapid, multilevel, thresholding technique based on high and low pass filters. The choice of a poor population can result in poor segmentation in multilevel thresholding [7].

The image segmentation of brain MRI is a significant and difficult problem dealing with grain mapping [8]. Correct categorization of magnetic resonance imagery by types of tissues of gray matter (GM), cerebrospinal fluid (CSF), and white matter (WM) at voxel level offers an approach to evaluate brain architecture. Among these, the quantization of WM and GM volumes has large significance for various neurodegenerative disorders; for example in movement disorders like Parkinson's disease and its associated syndromes, Alzheimer's disease, in inflammatory diseases, or WM metabolic in posttraumatic syndrome or congenital brain malformations [9].

However, the automatic image segmentation of brain magnetic resonance image (MRI) is still a persistent problem. Reliable and automatic categorization of tissues is further intricate due to the common characteristics among resonance intensities of various classes of tissues and by the presence of a spatially smooth altering intensity inhomogeneity. Hence, intensity-based algorithms which are fully automated demonstrate high sensitivity to a variety of noise artifacts, such as intertissue intensity and intratissue noise contrast reduction. Over the past few years, numerous algorithms have been proposed for image segmentation. In image segmentation, the popular method is the thresholding method owing to its efficiency and simplicity. If the target can be distinguished from the background, there will be a bimodal image of the histogram, after which it can easily reach the threshold simply by selecting the bottom of the valley as a threshold point. Nevertheless, in the majority of real images, no visibly noticeable marks between the background and the target are present. Image segmentation by using genetic algorithm methods has suggested a favorable threshold method to be extensively applied [10]. Using this method, the drawback can be overcome. However, calculating the optimum threshold requires a lengthy calculation time.

For resolving numerous optimization issues, the ACO algorithm has been successfully applied, even though it has a limited number of applications in the field of image processing. Recently, researchers have started to implement ACO algorithm to image processing problems, such as texture classification and edge detections [11−13].

In this chapter, we have adopted a novel approach by implementing the idea of ant colonies to the segmentation of iris and brain MRI images. Apart from Tabu search (TS), genetic algorithm (GA), artificial bee (AB), and simulated annealing (SA), ACO [14,15] is another special metaheuristic search algorithm that is useful in complex combinatorial optimization issues, for instance, graph-coloring issues, traveling salesman-based problems, vehicle routing issues, or quadratic assignment based problems, etc. The parallel and discretionary nature of ACO is appropriate for digital images. The purpose behind this is that ACO can search smartly and also possesses fine characteristics like positive feedback, distributed computation, and robustness [16].

In this research, image segmentation is viewed as delineating the area of pixels having a similar background texture. For each of the pixels present in an image, its gradient and brightness, collectively with the brightness and gradient of bordering pixels, are considered as local texture features [17]. In the case of ACO, artificial ants' movement is inclined by such local texture features, and the global pheromone (e.g., food trail secretions or pheromones) distribution on the image of a sizeable quantity of artificial ants tells the texture representation and region segmentation results. Following segmentation, texture representation outcomes are compared for few specific regions of iris and MRI images of patients suffering from iris infection or brain injury. The results obtained from experiments have proved the usefulness of ACO in the segmentation of images; moreover, they have also proved the discriminability of the texture representation based on ACO [18]. Thus the motivation behind this paper is to apply the ant colony approach to the image thresholding problems for medical images (e.g., iris and brain MRI images).

2.2 IMPLEMENTATION OF ANT COLONY OPTIMIZATION ALGORITHM

ACO-based algorithm was invented by an Italian researcher, M. Dorigo [19,20]. The algorithm was inspired by the examination of a real colony of ants and used for finding the best possible path to the source of food in the food-hunting process. In real life, ants are known as social insects living in colonies. Observations reveal that their behavior is targeted more on the survival of their colony as a whole instead of survival of an individual constituent of the colony. The key defining factors for ACO algorithms are discussed next.

2.2.1 Isula Framework

The Isula framework is a computer framework that permits a simple implementation of ACO algorithms using Java Programming Language. This framework includes the common elements that are present in the metaheuristic to permit algorithm designers to reuse common behaviors. Through Isula, optimization problems can be solved with ACO in a few code lines. For solving a problem via an ACO algorithm, we require a colony of agents (known as Ants), a graph demonstrating the problem statement, and a pheromone data structure for allowing communication between these agents.

The Isula framework provides the basic flow for execution of an algorithm in the ACO metaheuristic. We can use the implementations previously accessible for AntColony and AcoProblemSolver; however, we can also extend it based on our requirements. We need to create our Ant instance as per the requirement of our project; still, a lot of functionality is already available for base implementation. For reference, we can look at the projects in the section "Isula in action." Every algorithm of ACO has a set of custom-made behaviors which are executed throughout the solution processes; such behaviors can have a global effect (DaemonAction instances, such as rules of pheromone update) or only impact an ant and its solution (such as rules of component selection: regarded as subclasses of Ant Policy). Isula contains few such behaviors for some representative algorithms; however, we may need to define our policies or expand the ones that are available already (Fig. 2.1).

A significant and fascinating fact concerning ants is their foraging behavior, particularly, how they can discover the shortest path between food resources and their nest [22].

This section presents image segmentation process with ants having different feature use for hunting food resources. Additionally, the source of food is merely the best threshold of image segmentation. Fig. 2.2 illustrates a block diagram of the proposed approach for image segmentation.

We can describe the "food" in our proposed algorithm as the reference object which is remembered by ants during image segmentation

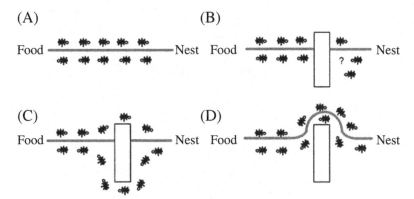

FIGURE 2.1 Ants' food searching approach: (A) ants traveling in a pheromone between food and their nest; (B) interruption of the trail due to an obstacle; (C) discovering two routes for trail movement; and (D) establishing the most favorable route [21].

phenomenon. For ease, we manually chose an r-radius locality $n_r(o)$ of a particular pixel o in the image. Subsequently, the food in the memory of ith ant for time $t = 0$ can be initialized as below:

$$f_{i,t=0} = n_r(o), \tag{2.1}$$

$$n_r(o) = \{e \in I | \quad ||e - o|| < r\}, \tag{2.2}$$

Here, r is an individual ant's state of being exhibited by its position; whereas, e denotes pixels that ants bypass through. I denote the pixels present in the image undergoing segmentation. When ants discover a new source of food, the food in their memory will be revitalized as per the rules of the transition.

After the food definition, the job of the ants is to find pixels having similar properties. For finding similar pixels, ants possess the capability to evaluate pixels to the particular reference food for which they have been searching.

The ants will move from one pixel to another at each step of the iteration. We suppose that the probability of transition of an ant also gets affected by the presence of other ants around it. The influence caused by other ants is limited to a given window, w. Subsequently, on the lattice, the normalized probability of transition to move from pixel to pixel at any time, t, is defined below:

$$p_i = \frac{w(\sigma_i)((v^* w \sigma_i) + E_i)}{\sum w(\sigma_i)((v^* w \sigma_i) + E_i)} \tag{2.3}$$

$$w(\sigma) = \left(1 + \frac{\sigma}{1 + \delta\sigma}\right)^\beta, \quad E = \frac{n_r - \text{food}}{nr}, \sigma$$

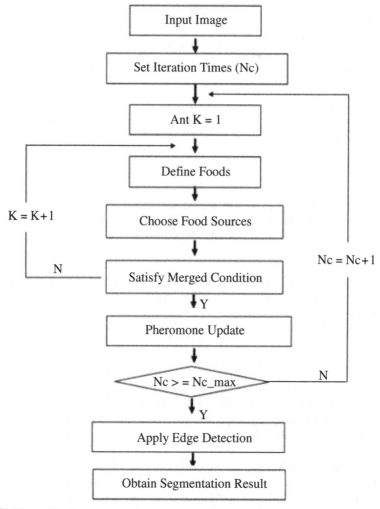

FIGURE 2.2 The flowchart of the proposed image segmentation algorithm.

Here, a pixel r has a pheromone density $\sigma(r)$, whereas, n_r − food is the amount of defined food, v denotes a pheromone rate of decay at each iteration, and nr is the quantity of the whole food. According to this definition, the movement of ants is more like a mass action, which can improve the ability of ants to find a food source.

After the value of p surpasses the threshold, an ant will then consider the target as a source of food. Throughout the process of searching for a source of food, every ant has the value of the threshold. Afterwards, we can define the pheromone deposition τ at pixel as follows:

$$\tau = \begin{cases} \eta & p_i < \lambda \\ \eta + p^*u & p_i \geq \lambda \end{cases} \tag{2.4}$$

In this equation, η is a constant quantity of pheromone; whereas, p is a constant weighting coefficient; and u is the rate at the ants looking for a source of food. Hence, the amount of pheromone changes on every path with the movement of the ant and is adjusted on every path through one circulation. According to Eq. (2.4), a new method has been adopted on updating the pheromone. Lastly, the general termination condition is to set the iterations number.

As individuals, ants are unsophisticated living beings and are unable to correspond or hunt effectively for food sources. However, they are intelligent when functioning in the form of a group, and successfully hunt and gather food for their colony. ACO techniques have been inspired by this collective intelligent behavior of ants. During ants' foraging process, they communicate via a chemical substance known as pheromone. While traveling, an ant deposits a constant quantity of pheromone that can be followed by several other ants. When searching for food, ants are usually inclined to follow trails with a higher concentration of pheromones [23]. Two main working operators in ACO algorithms are discussed in the next section.

2.2.2   Ant Route Construction

During the first stage, the traveling ants usually construct a route randomly on their path to food. Nonetheless, the ants succeeding chase the probability-based path construction scheme.

2.2.3   Ant Pheromone Update

Two important stages are involved in this process. Initially, a special chemical known as a pheromone is left on the path crossed by the individual ants. Later, this deposited pheromone evaporates. The amount of pheromone that is updated on an individual route is a cumulative outcome of these two stages.

2.3   IMAGE SEGMENTATION TECHNIQUES

For acquiring information from pictures through digital image analysis, the standard characteristics of images must be identified. This can be achieved by employing several characteristics of the contents of the images, such as edges, color strength, and regions. Techniques based on image content can be further divided into edge-based, thresholding-based, clustering-based, region-based, and grid-based methods.

2.3.1 Threshold-Based Segmentation

For portioning of the image, slicing techniques and histogram thresholding techniques are used. These techniques can be directly applied to an image; however, they can also be combined with post- and pre-processing techniques. Thresholding is possibly the most widely used technique for the segmentation of an image.

An algorithm that uses a histogram for probability values to define the optimal threshold value is one that maximizes the variance between classes, and was proposed by Otsu [24].

2.3.1.1 Otsu' Algorithm

1. Chose an initial approximate for T.
2. Segment the image through T. As a result, two groups of pixels will be produced: G_1 containing all pixels having values of intensity $> = T$; and G_2 containing all pixels having intensity values $<T$.
3. Calculate the average values of intensity $\mu1$ and $\mu2$ for the pixels present in regions G_1 and G_2, respectively.
4. Calculate a new value of the threshold.
5. Repeat steps 2−4 until the difference in the value of T in consecutive iterations becomes less than the predefined parameter T_0.

2.3.1.2 Ant Colony Optimization-Based Multilevel Thresholds Selection

The process of multilevel segmentation by using a genetic algorithm fails to provide sufficient results. To solve this problem through ACO, the basic steps of ACO are described in detail.

Ants exist in highly organized colonies in which they interact with others in perfect harmony. As stated earlier, the pheromone which ants leave on the ground while traveling decreases with time due to evaporation. Moreover, the amount of pheromone deposited by ants is dependent on the number of ants who use that trial. To find the shortest possible path between their colony and the source of food, ants adopt a unique collective organization technique which can be simulated to solve both static as well as dynamic optimization problems [25]. Ants can find the shortest path between their nest and food source without using visual information and, therefore, do not possess any global world model, and are adaptable to the environmental changes.

The probability of an ant choosing a particular path over another is presided by the quantity of pheromone on the prospective path of interest. As the quantity of pheromone on a path evaporates with time, the ants adopting the shorter path will return first to their nest with food. The most pheromones are present in shorter routes since the paths have fresh pheromones which have not yet evaporated and, hence, will be more attractive for ants returning to the source of food. This possibility (even though small) allows other trails to evaporate which is helpful as it allows discovery of alternate, or shorter, pathways or fresh food sources. As the pheromone trail evaporates with the passage of time, the trail will become less detectable. Hence, the longer trails are less attractive which benefits the ant colony.

2.3.1.3 Algorithm for Ant Colony Optimization

1. Ants travel around the colony to find sources of food.
2. After finding a source, the ant returns back to its nest.
3. During traversing, ants leave a trail of pheromones.
4. The follower ants of the first ant go after the pheromones deposited by the first ant.
5. As a result of this transaction, the deposition of the pheromone on the trail will be strengthened.
6. The quantity of pheromones in each traversal will evaporate.
7. If there are two paths to get to the same source of food, the ant finds the shortest path between their nest and food with the help of the fresh pheromones.

2.3.2 Edge-Based Segmentation

Edge detection is a technique used to recognize and locate severe discontinuities in an image. This technique represents the detected edges in images as the object boundaries which are utilized to recognize the objects. There are several ways to execute edge detection using different types of detectors, for example, Canny, ACO, Robert, Sobel, Prewitt, etc. However, an ACO-based edge detection technique has been investigated, which will be discussed next.

2.3.2.1 Ant Colony Optimization-Based Edge Detection Initialization

The initialization of parameters α and β takes place at the beginning of the experiment. The probable heuristic data collected. The calculation of some ants is done as $K:\sqrt{M_1 \cdot M_2}$ where M_2 is the width and M_1 is the image length. All the K number of ants are spread on a 2D image such that almost one ant is present on each pixel. Each image pixel contains a node and an initial pheromone matrix value, that is, a constant value is assigned to τ^0 [26].

2.3.2.2 Ant Colony Optimization-Based Structuring Process

By picking from a cluster of K ants at each construction step, one of the ants is moved L steps on the image (say I). The ant (A_k) supposedly moves from a node (l, m) to a neighboring node (i, j) by the probabilistic transition matrix as defined in Eq. (2.5).

$$P^n_{(l,m)(i,j)} = \frac{\left(\tau^{n-1}_{i,j}\right)^\alpha \cdot \left(\eta_{i,j}\right)^\beta}{\sum_{(i,j)\varepsilon\Omega_{(l,m)}}\left(\tau^{n-1}_{i,j}\right)^\alpha \cdot \left(\eta_{i,j}\right)^\beta} \tag{2.5}$$

where $\tau^{n-1}_{i,j}$ denotes the value of a pheromone at a particular node (i, j) and it has been established the neighboring node of a (l, m) node is $\Omega_{(l,m)}$. In other words, $\Omega_{(l,m)}$ represents all the pixels which can be present in the 8-i, j neighborhood of a pixel positioned at (l, m). The metaheuristic data of a node (i, j) is represented by η. To quantify the heuristic data [26,27], the configuration which is typically local at each level (i.e., i, j) is defined as:

$$\eta_{i,j} = \frac{1}{Z} V_{c(I_{i,j})} \tag{2.6}$$

where Z is known as the normalization factor utilized to isolate error and is defined as:

$$Z = \sum 1{:}M_1 \sum 1{:}M_2 V_{c(I_{i,j})} \tag{2.7}$$

For Eq. (2.7), $I_{i,j}$ signifies the intensity value a pixel (i, j) present in the image (I).

The disparity in the intensity values of an image's pixels typically depends on "c" which is formed by a group of similar pixels. It has been established that the group of pixels constitute the function $V_{c(I_{i,j})}$.

The function $V_c(I_{i,j})$ mainly depends on its neighboring pixel groups (c), which is defined in Eq. (2.8).

$$V_c(I_{i,j}) = f(|I_{i-2,j-1} - I_{i+2,j+1}| + |I_{i-2,j+1} - I_{i+2,j-1}| +$$
$$|I_{i-1,j-2} - I_{i+1,j+2}| + |I_{i-1,j-1} - I_{i+1,j+1}| + |I_{i-1,j} - \tag{2.8}$$
$$I_{i-1,j} + I_{i-1,j+1} - I_{i-1,j-1} + I_{i-1,j+2} - I_{i-1,j-2} + I_{i,j-1} - I_{i,j+1})$$

Eq. (2.8) provides insight into the function which guarantees that the likelihood of the sharp-turn shapes in the image is less than that of the small-angle turns. Therefore every ant in a particular colony has the propensity to travel in the forward direction.

Modeling of Eq. (2.8) for bringing out changes in the respective shapes can be performed as follows:

$$f(x) = \lambda x \text{ for } x \geq 0 \tag{2.9}$$

$$f(x) = \lambda x^2 \text{ for } x \geq 0 \tag{2.10}$$

$$f(x) = \sin\left(\frac{\pi x}{2\lambda}\right) \quad 0 < {=} x < {=} \lambda \tag{2.11}$$

$$\text{else } f(x) = \frac{\pi x \sin\left(\frac{\pi x}{\lambda}\right)}{\lambda} \tag{2.12}$$

The parameter λ incorporated in these equation functions indicates that this function is responsible for adjusting the respective shapes of each function.

2.3.2.3 Ant Colony Optimization-Based Updating Process

This section proactively emphasizes the updating process which involves the steps moved through by each ant after each neighboring ant, followed by all ants on each construction step. An effort has been made to alter only a single updating process in the ACO algorithm to yield a binary image with missing data [28].

After the movement of each ant, the updating process updates the pheromone matrix which is given in Eq. (2.13).

$$\tau_{i,j}^{n-1} = \frac{(1-\rho), \tau_{i,j}^{n-1} + \rho, \ \Delta_{i,j}^k \text{ if } (i,j) \text{ is visited by } k\text{th ant}}{\tau_{i,j}^{n-1}, \ \text{ otherwise}} \tag{2.13}$$

where ρ is signifies the evaporation rate and a heuristic matrix is used to determine $\Delta_{i,j}^k$, that is, $\Delta_{i,j}^k = \eta_{i,j}$

The heuristic data from a heuristic matrix is gradually added into the memory of the ants. The second update is carried out at the conclusion of each building step (after the movement of all the (K) ants within the construction step. The movement of all the ants at the conclusion of the construction step yields Eq. (2.14).

$$\tau^n = (1-\psi), \ \tau^{n-1} + \psi\tau^n \tag{2.14}$$

where ψ is known as the pheromone decay coefficient.

The step demonstrated in Eq. (2.14) signifies the pheromone matrix which is updated in this particular instance with the consideration of the pheromone matrix and the decay coefficient.

2.3.2.4 Decision Process

A threshold value (i.e., T) is typically used on τ^N (pheromone matrix) to obtain the edge information for a particular image. The iterative method has been proposed by various researchers for computing the threshold in ant ACO algorithms. It has been reported that a normalized intensity value (typically in a range of [0,1]) is adapted to carry out the conversion of the intensity image to a binary image [29]. Concerning the initial threshold value, the segmentation of the histogram into two distinct parts takes place. Computation of the mean values (for gray values) related to the foreground pixels along with sample mean (for the gray values) related to the pixels' background is carried out. Therefore this current threshold value can be regarded as the average value of the two specimens. The starting threshold value (T^0) is ultimately considered as the pheromone matrix's mean value. Every index value in the pheromone matrix is signified as either above the initial threshold value or below the threshold value. Based on these two classifications, the average of all the mean values is computed that is considered as a (new) fresh threshold value. This averaging process is carried out until the threshold value approaches a constant.

Step 1: Initialization of $T^{(0)}$ as given in Eq. (2.15)

$$T^{(0)} = \frac{\Sigma_{i=1:M_1} \Sigma_{j=1:M_2} \tau_{i,j}^{(N)}}{M_1 M_2} \tag{2.15}$$

Step 2: Separation of $\tau^{(N)}$ (pheromone matrix) into two separate classes using $T^{(l)}$ in which one of the classes contains entries of τ with values less than the threshold value of $T^{(l)}$, while the other class contains the remaining entries of τ. Concerning these categories, the mean calculation for the above categories is carried out as follows:

$$m_L^{(l)} = \frac{\Sigma_{i=1:M_1} \Sigma_{j=1:M_2} g_T^L(l)(\tau_{i,j}^{(N)})}{\Sigma_{i=1:M_1} \Sigma_{j=1:M_2} h_T^L(l)(\tau_{i,j}^{(N)})} \tag{2.16}$$

$$m_U^{(l)} = \frac{\Sigma_{i=1:M_1} \Sigma_{j=1:M_2} g_T^U(l)(\tau_{i,j}^{(N)})}{\Sigma_{i=1:M_1} \Sigma_{j=1:M_2} h_T^U(l)(\tau_{i,j}^{(N)})} \tag{2.17}$$

where

$$g_{T^{(l)}}^L(x) = \{x, \text{if } x \leq T^{(l)} \text{ otherwise } 0 \tag{2.18}$$

$$h_{T^{(l)}}^L(x) = \{x, \text{if } x \leq T^{(l)} \text{ otherwise } 0 \tag{2.19}$$

$$g_{T^{(l)}}^U(x) = \{x, \text{if } x \leq T^{(l)} \text{ otherwise } 0 \tag{2.20}$$

$$h_{T^{(l)}}^U(x) = \{x, \text{if } x \leq T^{(l)} \text{ otherwise } 0 \tag{2.21}$$

Step 3: Iteration index is maintained as $l = 1 + 1$ which ultimately updates the threshold as given in Eq. (2.22).

$$T^{(l)} = \frac{m_L^{(l)} + m_U^{(l)}}{2} \tag{2.22}$$

Step 4: The new threshold value $(T^{(l)})$ is assessed for similarity with $T^{(n-1)}$. If the similarity index is zero, then step 2 is repeated; otherwise the procedure is ended and the threshold value is documented.

Consequently, the determination of the definite edge $(E_{i,j})$ in an image is carried out using the pixel (i, j).

$$E_{i,j} = \begin{cases} 1 \text{ if } \tau_{i,j}^{(N)} \geq T^{(l)} \\ \overline{} \\ 0 \end{cases} \tag{2.23}$$

where $E_{i,j}$ denotes the image obtained at pixel (i, j)

A variety of parameters and their values employed in the above calculations are listed below:

1. $\alpha = 1$, that is, weighting factor for information of pheromone matrix.
2. $\beta = 0.1$, that is, weighting factor for information of heuristic matrix.
3. $N = 4$, that is, sum of all construction processes.
4. $\Psi = 0.05$, that is, pheromone decay coefficient.
5. $L = 40$, that is, movement steps of an ant.
6. $\lambda = 1$, that is, the adjustment factor.
7. $\rho = 0.1$, that is, evaporation rate.
8. $\Omega = 8$, that is, connectivity with neighborhood.
9. $\tau_{\text{init.}} = 0.0001$, that is, the primary value of every constituent in pheromone matrix.
10. $K = \sqrt{M_1 M_2}$, that is, number of the total ants (the image size).

2.4 EVALUATION OF SEGMENTATION TECHNIQUES

Assessment is essential in deciding which segmentation algorithm is suitable to be selected. For extraction of the boundary, the edge detection is assessed by using root-mean-square-error (RMSE), mean-square error (MSE), peak signal-to-noise ratio (PSNR), and signal-to-noise ratio (SNR) [30–32].

2.4.1 Mean-Square Error

MSE value denotes the average difference of the pixels all over the image. A higher value of MSE designates a greater difference amid the original image and processed image. Nonetheless, it is indispensable to be extremely careful with the edges. The following equation provides a formula for calculation of the MSE.

$$\text{MSE} = \frac{1}{N} \sum \sum (F_{ij} - o_{ij})^2 \tag{2.24}$$

here, N is the image size, O is the original image, whereas, E is the edge image.

2.4.2 Root-Mean-Square-Error

The RMSE is extensively used for measuring the differences between values forecasted by an estimator, or model, and the values that are observed. The RMSE aggregates the magnitudes of the predictions errors several times into a single distinct measure of predictive power. It is a measure of precision. The RMSE value can be calculated by taking the square root of MSE.

$$\text{RMSE} = \sqrt{\frac{1}{N} \sum \sum (E_{ij} - o_{ij})^2} \tag{2.25}$$

2.4.3 Signal-to-Noise Ratio

SNR describes the total noise present in the output edge detected in an image, in comparison to the noise in the original signal level. SNR is a quality metric and presents a rough calculation of the possibility of false switching; it serves as a mean to compare the relative performance of different implementations [33–35]. SNR is estimated by Eq. (2.26):

$$\text{SNR} = \left[\frac{\Sigma_i \Sigma_j (E_{ij})^2}{\Sigma_i \Sigma_j (E_{ij} - o_{ij})^2} \right] \tag{2.26}$$

2.4.4 Peak Signal-to-Noise Ratio

The PSNR calculates the PSNR ratio in decibels amid two images. We often use this ratio as a measurement of quality between the original image and the resultant image. The higher the value of PSNR, the better will be the quality of the output image. For calculating the PSNR, MSE is used. We calculate the PSNR by using Eq. (2.27).

$$
\begin{aligned}
PSNR &= 10 \cdot \log_{10} \left(\frac{MAX_I^2}{MSE} \right) \\
&= 20 \cdot \log_{10} \left(\frac{MAX_I}{\sqrt{MSE}} \right) \\
&= 20 \cdot \log_{10}(MAX_I) - 10 \cdot \log_{10}(MSE)
\end{aligned}
\tag{2.27}
$$

Here, MAX_I is the maximum possible pixel value of the image

2.5 EXPERIMENTS AND RESULTS

Boundary extraction was initially carried out in image segmentation process. The operators of edge detection were employed on the real iris and brain MRI images, while some of the test experiments were also performed on images obtained from different image datasets, for example, Berkley dataset, Caltech dataset, etc. [36,37]. The experimental results exhibit the visual output of the edge detection. Two types of images (brain MRI and iris) are experimentally shown next.

2.5.1   Ant Colony Optimization-Image-Segmentation Using the Isula Framework

A Java Program (for ACO-image-segmentation) was implemented to identify image segments using the ACO algorithm. Segmentation is a two-phase process. Initially, the image (brain MRI/iris) is extracted using the Ant Colony algorithm followed by the application of a clustering procedure. The ACO algorithm for extraction of image segments is centered on Max−Min Ant System which is implemented using the Isula Framework.

The implemented ACO process has the following characteristics:

1. The solution components are Clustered Pixels, that is, each image pixel (of brain MRI or iris image) is assigned to one of the three clusters.
2. The available Ant class in the Isula framework was extended to back the image-clustering class.
3. The ants used for the particular types of problems possessed memory concerning the current position in an image along with the Map for storage of Cluster.
4. The determination of the Solution Quality is quite expensive; therefore this was incorporated as an instance variable.
5. These Ants were constructed so that they only considered certain pixels (cerebrum or iris) in the images. That data was provided by the earlier binary thresholding algorithms.
6. The Max−Min Ant Systems' policies accessible in Isula were reprocessed with minor tailoring. These policies include node selection and pheromone update.

Boundary extraction was initially carried out in image segmentation process. The operators of edge detection were employed on the images obtained from a segmentation image dataset, for example, Berkley dataset, Caltech dataset, Corel dataset, etc. The experimental results exhibit the visual output of the edge detection. Two types of images (brain MRI and iris) are experimentally shown next.

2.5.2   Performance Testing Ant Colony Optimization Image Segmentation Algorithm

Various test images were initially used to explore the ACO Image Segmentation Algorithm's (i.e., ACO-ISA) performance. The key parameters were given following values throughout the experiment.

A total number of ants was around 20% of the sum of pixels on a particular image, that is, $\eta = 0.3$; $K = 0.985$, while the maximum number of iterations was 200. These parametric values seem to be suitable for approximately all of the images, and can be altered with less effective (minor) difference in the resulting quality [38−40]. However, the other experimental parameters were selected in accordance with different image types, because of their close association with the histogram of various types of images. Examples of the performance of ACO segmentation along with its comparison with conventional edge detection techniques are illustrated in Figs. 2.5 and 2.6.

2.5.3 Application of Ant Colony Optimization on Segmentation of Brain MRI

The performance of the ACO algorithm was assessed by its implementation on the real brain MRI. Processing of the tests was performed on the high-speed workstation. Fig. 2.3 displays the original brain MR images (256×256) which are 8-bit grayscale having the intensity values in the range of 0−255. The experimental study offered proof of the efficacy of the proposed method. The results are analyzed in the following section.

The results of image segmentation carried out using the indigenously developed algorithm for a different set of iterations are shown in Fig. 2.4. Considering the time of iterations, the segmentation of the WM takes place conveniently after 200 iterations for a T_1-weighted image. Additionally, the segmentation of the GM also takes place easily after 200 iterations for a T_2-weighted image. The same MRI image was investigated using the metaheuristic techniques for image segmentation which included the algorithms based on canny, artificial bee, Robert, Sobel, and Prewitt detectors. Figs. 2.4 and 2.5 exhibit the comparison of the image segmentation results concerning the number of iterations and type of detectors. It can be observed that our proposed approach (using an ACO algorithm) provides more accurate results than other conventional techniques.

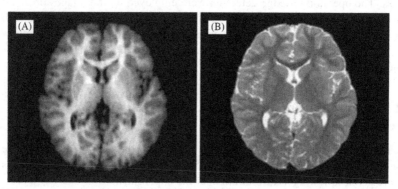

FIGURE 2.3 The original brain MRI images. (A) Skull-stripped (or T_1)-weighted image; (B) skull-stripped (or T_2)-weighted image.

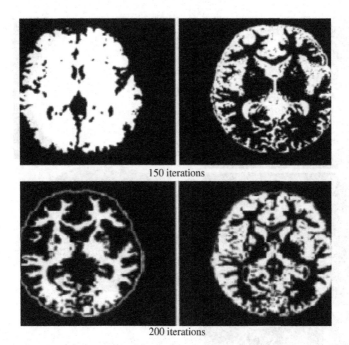

150 iterations

200 iterations

FIGURE 2.4 The image segmentation results obtained at different iterations utilizing the ACO algorithm. In both cases, the first column illustrates the T_1 image results while the second column demonstrates the T_2 image results.

2.5.4 Ant Colony Optimization-Image Segmentation on Iris Images

The performance of an ACO-based image segmentation algorithm for iris images is demonstrated in Fig. 2.6. It can be easily observed that the pigmentation and collarette regions are distinct under the ACO-ISA based detector. It also establishes that the ACO-ISA processed image depicts better quality for further application of region edging techniques.

In our experimental processes, the image segmentation ability of ACO-ISA was investigated on iris areas of three types of typical textures, that is, normal texture, the radii Solaris, and the hyperpigmentation textures. After the processing of ACO-IRIS and ACO-TRA (texture representation ability, the alignment of each region to 3×3 sub-zones took place followed by the calculation of histogram for each sub-zone. The comparison of two regions was made by taking the mean of histogram distances for all corresponding pairs of sub-zones. A sum of 900 sample regions was involved for each kind of the texture. The average distances (interclass distance) are demonstrated in Table 2.1. It is obvious from the results that different texture regions are well-distinguished for the ACO segmentation technique compared to the

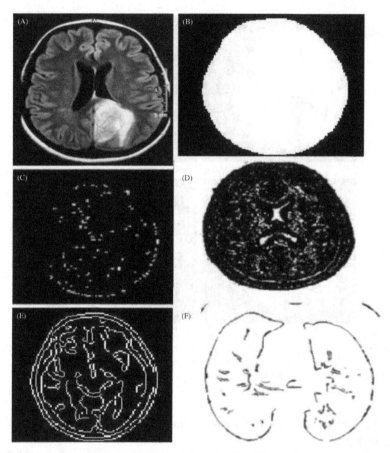

FIGURE 2.5 Brain MRI edge detection information for different image segmentation and image edge detection methods. (A) Original image of brain MRI. (B) Sobel edge detector's performance (threshold = 0.55; iterations = 200). (C) The Prewitt edge detector's performance (threshold = 0.26; iterations = 200). (D) The Robert edge detector's performance (threshold = 0.19; iterations = 200). (E) Canny edge detector's performance (threshold = 0.55; iterations = 200). (F) ACO edge detector's performance (threshold = 0.55; iterations = 200).

results obtained from other techniques. Additionally, we also designed a self-organized feature map (SOFM) neural network a tool for texture recognition, and around 95.1% accurate recognition rate was attained. This unveiled that our proposed technique has the potential to be useful in automatic disease diagnosis systems using different types of images (iris, MRI, etc.).

To scrutinize the performance of the ACO algorithm, real iris and brain MRI images were used in the experiments. The output results obtained from traditional edge detection techniques (Figs. 2.5B−F and 2.6B−E) exhibit the results with 200 iterations and the parameters were adjusted to $\beta = 3.5$, $\eta = 0.08$, $v = 0.017$, $p = 1.3$; while many ants were 10% of all the pixels.

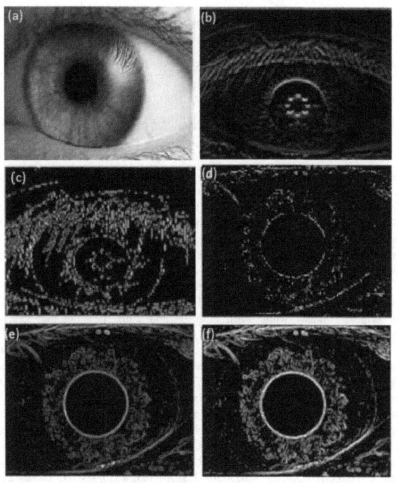

FIGURE 2.6   (A) Original iris image. (B) Performance of Sobel detector (at 200 iterations). (C) Performance of Canny detector at 200 iterations. (D) Performance of Robert edge detection at 200 iterations. (E) Performance of Prewitt detector at 200 iterations. (F) Performance of ACO detector at 200 iterations.

TABLE 2.1 Interclass Spaces Between Different Iris Regions

Interclass Distance	Radii Solaris	Hyperpigmentation	Normal Region
Hyperpigmentation	13.67	–	84.53
Normal region	97.95	84.47	–
Radii Solaris	–	13.62	97.87

TABLE 2.2 Time Passed (in Seconds) by Traditional Edge Detection Techniques as Compared to ACO

Image Type	Canny	Robert	Sobel	Prewitt	Artificial Bee	ACO
Brain MRI	0.285	0.868	0.084	0.734	0.688	114.75
IRIS	2.67	0.085	0.078	0.067	0.078	60.08

Higher values of v and smaller values of β lead to higher probability of probing new paths. Therefore for the deposition of further pheromones in a digital habitat, the number of iterations must be increased. Experimental results for the ACO-based technique showed promising results. The process times of ACO has been improved (shortened) compared to the processing times of ACO employed in former research [41]. However, ACO still requires more time than conventional edge detection techniques, as demonstrated in Table 2.2.

Our current experiment involves the experimentation of five different images (five each of iris and brain MRI images). Figs. 2.5A and 2.6A display the original images while Figs. 2.5B−F and 2.6B−E show the results obtained from conventional algorithms (Canny, Robert, Prewitt, Sobel, etc.). Both of the image sets processed with the ACO technique exhibited superior performance compared to the images processed with conventional algorithms regarding efficiency and effectiveness. To validate the effectiveness of ACO results, experiments were conducted on various other (real and synthetic) images. The segmentation (using images other than iris and brain) results were filtered by the Robert, Canny, Sobel, Prewitt, and ACO edge detectors using the Java framework in the MATLAB toolbox. Number of ants in digital habitat were 10% of the total number of pixels and essential parameters were $\delta = 0.03$, $\beta = 3.5$, $v = 0.017$, $p = 1.3$, $\lambda = 0.5$, and $f_i = 0.09$. Tables 2.3 and 2.4 depict the difference in performance for various edge detectors.

2.5.5 Comparison of Results

ACO metaheuristics are specialized in solving the optimization problems in every iteration step by vaporizing and updating the pheromone density. In Figs. 2.4−2.6 ACO image segmentation with a different number of iterations using different techniques are illustrated. Fig. 2.4 depicts the brain MRI images segmented with 100 and 200 iterations using the ACO technique. Increasing the number of iterations improves the segmentation performance of ACO using a lower number of ants, as depicted in Fig. 2.4. The decrease in some ants to 1% of the total number of pixels and the increase in the number of iterations from 20 to 200 results in significantly improved segmented images with a higher probability of edge detection.

TABLE 2.3 Edge Operator (Detector) Evaluation Table for Brain MRI Image

	Sobel	Prewitt	Robert	Artificial Bee	Canny	ACO
RMSE	5.5760	5.5826	5.9294	4.6394	4.6394	3.3519
SNR	43.0944	42.8559	29.2034	68.2989	68.2989	84.3721
PSNR	33.2047	33.1944	32.6719	34.8026	34.8026	37.6483

TABLE 2.4 Edge Operator (Detector) Evaluation Table for IRIS Image

	Sobel	Prewitt	Robert	Artificial Bee	Canny	ACO
RMSE	2.2728	2.2774	2.2940	2.1175	2.1175	1.8119
SNR	43.8058	43.4136	42.2534	55.4086	55.4086	67.5980
PSNR	41.0000	40.9827	40.9195	41.6154	41.6154	43.7026

The comparison of ACO performance with conventional edge detection techniques is shown in Figs. 2.5 and 2.6, respectively. Fig. 2.5 shows the comparison of brain MRI image segmentation using algorithms of Canny, Artificial Bee, Sobel, Robert, Prewitt, and ACO techniques. The traditional edge detection techniques show unclear edges with poor segmentation. However, the results of the ACO technique depict the compensation of the results regarding edge detection for suitable image segmentation [17].

The ACO method utilized two thresholds to identify weak and strong edges. It only incorporated those weak edges in the output which were linked to strong edges. Consequently, the technique is accurate and robust to noise with higher probability of detecting the true weak/strong edges.

Additionally, the proposed method exploits the movement of a definite number of ants on the image based on the localized disparity in the image's intensity value. The localized disparity in the image's intensity value is used to develop a pheromone matrix, which provides the edge data of the image.

The statistical comparison of the ACO technique with other traditional techniques is carried out using the operator parameters which include RMSE, SNR, and PSNR. Tables 2.3 and 2.4 show the comparative values of the operator parameters for five iris and five brain MRI images evaluated through six algorithms (Canny, Sobel, Robert, Artificial Bee, Prewitt, and ACO). RMSE provides the difference between the values of the edge-detected pixels and the original pixels for a particular technique. Low RMSE signifies less difference between the values of the original and processed images, which indicates the accuracy of the output image. SNR is a relative value of the signal concerning noise and must be high for the accuracy of the output. On the other hand, the PSNR peak should be higher for enhanced quality.

Table 2.3 shows the comparative results for RMSE, SNR, and PSNR values obtained by different detector types (algorithms) used on MRI images. Although the ACO algorithm offers more clear edges, a comparative study of different edge detection techniques was carried out for validating the boundary extraction of the images. Table 2.3 shows that RMSE for ACO is at a minimum (3.7519) which indicates higher accuracy. On the other hand, comparatively higher values of SNR and PSNR (82.3721 and 36.6483, respectively) depict higher image quality for ACO in MRI images.

Table 2.4 shows the comparative results for RMSE, SNR, and PSNR values obtained by different detector types (Canny, Robert, Sobel, Prewitt, ACO, etc., algorithms) used on iris images. Although the ACO algorithm offers more clear edges, a comparative study of the different edge detection techniques was carried out for validating the boundary extraction of the images. Table 2.4 shows that RMSE for ACO is at a minimum (1.9119) which indicates higher accuracy. On the other hand, comparatively higher values of SNR and PSNR (65.4980 and 42.5026, respectively) depict higher image quality for ACO in the iris images.

The following figures display the graphs showing the comparison of RMSE, SNR, and PSNR values for 10 (5 iris and 5 brain MRI) images.

Fig. 2.7 shows a graph for depicting the RMSE performance of different algorithms using five iris and five brain MRI images. It can be observed that ACO peaks are smallest among all, which indicates lowest RMSE. On the other hand, Robert and Sobel's detectors exhibit highest RMSE values, which indicate lower accuracy in image segmentation. Similarly, Figs. 2.8 and 2.9 show the SNR and PSNR for different algorithms applied on iris and brain MRI images. Both graphs display higher ACO peaks, which indicate higher

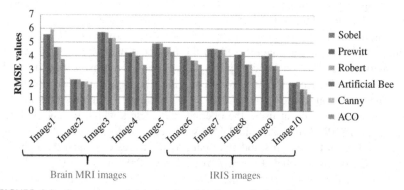

FIGURE 2.7 Comparison of image edge detection methods using root-mean-square-error (RMSE).

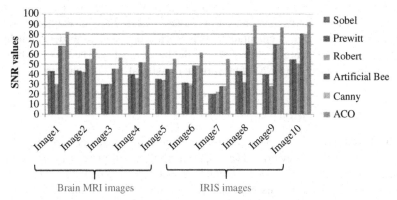

FIGURE 2.8 Comparison of image edge detection methods using signal-to-noise-ratio (SNR).

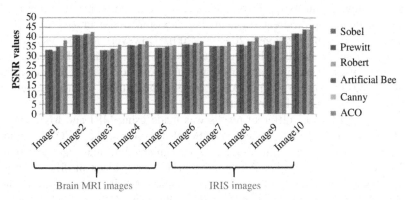

FIGURE 2.9 Comparison of image edge detection methods using peak-signal-to-noise ratio (PSNR).

segmentation and edge detection quality for the ACO-based technique. On the other hand, Robert and Prewitt's techniques show lowest peaks, which indicate their poor segmentation characteristics.

An adaptive edge detector is essential for obtaining a robust solution which is adaptable to the fluctuating noise levels of the images and helps to distinguish valid contents of images from visual defects induced by noise. The ACO's performance heavily depends on the changeable parameters (e.g., σ, typically known as the standard deviation of the Gaussian filter) and threshold values (e.g., T_1 and T_2). σ also regulates the Gaussian filter's size. Higher values of σ result in larger sizes of the Gaussian filter. This phenomenon implies higher and more blurring levels, essential for noisy images along with detection of the larger edges. The results and graphs shown in this chapter depict the higher effectiveness of the indigenously modified ACO algorithm proposed. The tuning of the parameters has yielded excellent results (better than the previous results quoted in the literature) concerning lower RMSE and higher SNR and PSNR, as indicated in Tables 2.3 and 2.4. Our main aim was to improve the image segmentation with enhanced image edge detection by lowering the error margin and enhancing the image quality. The task has been accomplished using the Java framework and modified ACO algorithm.

2.6 CONCLUSION

This chapter presents a novel ACO-based algorithm for image segmentation using the Java framework. The ACO method is inspired by the food-searching mechanisms in ant colonies. Experimental parameters for ACO were altered and the results were compared with the conventional segmentation and edge detection techniques. The study was performed on real iris and brain MRI images by the application of different algorithms (Canny, Robert, Sobel, Prewitt, and ACO). Five images each of iris and brain MRI were used to investigate six different segmentation algorithms. The images were iterated at different iterations (20−100−200). However, all the techniques exhibited better results at higher iterations (∼200). On the other hand, the number of ants was increased at lower iterations to observe the effect of increasing the number of ants. An increase in the number of ants resulted in better performance at lower iterations.

A comparison of different results was made using MRI and iris images. The results unveil that unlike conventional algorithms, the ACO algorithm offers superior results concerning image segmentation and edge detection. The images shown in Figs. 2.4−2.6 exhibit the comparison of different segmentation techniques (with different edge detectors). It is obvious from the images that the ACO algorithm provides better edge detection and image as compared to other techniques. Moreover, key operator parameters, such as RMSE, SNR, and PSNR, were also analyzed for all six algorithms. Tables 2.3 and 2.4 and Figs. 2.7−2.9 show the superior and improved performance of the ACO-based algorithm over conventional techniques. The basic drawback of the ACO-based detector is its higher time consumption during the algorithm application. The alteration in parameters has significantly enhanced the performance of the ACO technique by reducing the process time up to 50% of the times achieved previously.

The robustness of the ACO technique against image noise was investigated using real brain MRI and iris images. Experimental results exhibited efficient (significantly superior) performance of the ACO approach in comparison with traditional edge detection approaches. This chapter presents the improved results for solving image segmentation and edge detection problems using a modified (upgraded) ACO algorithm which yields superior results than those offered by the traditional approaches, that is, Canny, Prewitt, Sobel, Robert, etc.

REFERENCES

[1] S. Paul, S. Datta, S. Das, Rough-fuzzy collaborative multi-level image thresholding: a differential evolution approach, Mendel 2015, Springer, Cham, 2015, pp. 329–341.

[2] A. Singla, S. Patra, A fast automatic optimal threshold selection technique for image segmentation, Signal Image Video Process. 11 (2) (2017) 243–250.

[3] E.H. Guerrout, S. Ait-Aoudia, D. Michelucci, R. Mahiou, Hidden Markov random field model and BFGS algorithm for brain image segmentation, Proceedings of the Mediterranean Conference on Pattern Recognition and Artificial Intelligence, ACM, New York, NY, 2016, November, pp. 7–11.

[4] X. Li, L. Lu, L. Liu, G. Li, X. Guan, Cooperative spectrum sensing based on an efficient adaptive artificial bee colony algorithm, Soft. Comput. 19 (3) (2015) 597–607.

[5] C.H. Tung, Z.L. Wu, Binarization of uneven-lighting image by maximizing boundary connectivity, J. Stat. Manage. Syst. 20 (2) (2017) 175–196.

[6] A. Mukhopadhyay, U. Maulik, S. Bandyopadhyay, A survey of multiobjective evolutionary clustering, ACM Comput. Surv. (CSUR) 47 (4) (2015) 61.

[7] M.G. Pérez, A.S. Calle, A.B. Moreno, E. Nunes, V. Andaluz, A multi-level thresholding method based on histogram derivatives for accurate brain MRI segmentation, Rev. Politéc. 35 (2) (2015) 82.

[8] B. Minnaert, D. Martens, M. De Backer, B. Baesens, To tune or not to tune: rule evaluation for metaheuristic-based sequential covering algorithms, Data Min. Knowl. Discov. 29 (1) (2015) 237–272.

[9] I.A. Hodashinsky, D.Y. Minina, K.S. Sarin, Identification of the parameters of fuzzy approximators and classifiers based on the cuckoo search algorithm, Optoelectron. Instrum. Data Process. 51 (3) (2015) 234–240.

[10] S.L. Jui, S. Zhang, W. Xiong, F. Yu, M. Fu, D. Wang, et al., Brain MRI tumor segmentation with 3D intracranial structure deformation features, IEEE Intell. Syst. 31 (2) (2016) 66–76.

[11] M. Brajović, V. Popović-Bugarin, I. Djurović, S. Djukanović, Post-processing of time-frequency representations in instantaneous frequency estimation based on ant colony optimization, Signal. Process. 138 (2017) 195–210.

[12] K. Singh, R. Kapoor, S.K. Sinha, Enhancement of low exposure images via recursive histogram equalization algorithms, Optik: Int. J. Light Electron Opt. 126 (20) (2015) 2619–2625.

[13] S.H. Lim, N.A.M. Isa, C.H. Ooi, K.K.V. Toh, A new histogram equalization method for digital image enhancement and brightness preservation, Sig. Image Video Process. 9 (3) (2015) 675–689.

[14] M. Mavrovouniotis, S. Yang, Training neural networks with ant colony optimization algorithms for pattern classification, Soft. Comput. 19 (6) (2015) 1511–1522.

[15] M.M.T. Alobaedy, Hybrid ant colony system algorithm for static and dynamic job scheduling in grid computing, Doctoral dissertation, Universiti Utara Malaysia, 2015.

[16] G.G. Wang, A.H. Gandomi, X.S. Yang, A.H. Alavi, A new hybrid method based on krill herd and cuckoo search for global optimisation tasks, Int. J. Bio-Inspir. Comput. 8 (5) (2016) 286–299.

[17] T. Sağ, M. Çunkaş, Color image segmentation based on multiobjective artificial bee colony optimization, Appl. Soft. Comput. 34 (2015) 389–401.

[18] G. Greenfield, P. Machado, Ant-and ant-colony-inspired alife visual art, Artif. Life 21 (3) (2015) 293–306.

[19] M. Salari, M. Reihaneh, M.S. Sabbagh, Combining ant colony optimization algorithm and dynamic programming technique for solving the covering salesman problem, Comput. Ind. Eng. 83 (2015) 244–251.

[20] E.G. Talbi, Combining metaheuristics with mathematical programming, constraint programming and machine learning, Ann. Oper. Res. 240 (1) (2016) 171–215.

[21] T.X. Lin, H.H. Chang, Medical image registration based on an improved ant colony optimization algorithm, Int. J. Pharma Med. Biol. Sci. 5 (1) (2016) 17.

[22] M.D. Toksari, A hybrid algorithm of ant colony optimization (ACO) and iterated local search (ILS) for estimating electricity domestic consumption: case of Turkey, Int. J. Electr. Power Energy Syst. 78 (2016) 776–782.

[23] I. Bloch, Fuzzy sets for image processing and understanding, Fuzzy Sets Syst. 281 (2015) 280–291.

[24] A. Nayyar, R. Singh, Simulation and performance comparison of ant colony optimization (ACO) routing protocol with AODV, DSDV, DSR routing protocols of wireless sensor networks using NS-2 simulator, Am. J. Intel. Syst. 7 (1) (2017) 19–30.

[25] P. Shao, W. Shi, P. He, M. Hao, X. Zhang, Novel approach to unsupervised change detection based on a robust semi-supervised FCM clustering algorithm, Remote Sens. 8 (3) (2016) 264.

[26] T.M. Tuan, A cooperative semi-supervised fuzzy clustering framework for dental X-ray image segmentation, Expert Syst. Appl. 46 (2016) 380–393.

[27] M.A.J. Ghasab, S. Khamis, F. Mohammad, H.J. Fariman, Feature decision-making ant colony optimization system for an automated recognition of plant species, Expert Syst. Appl. 42 (5) (2015) 2361–2370.

[28] Y. Khaluf, S. Gullipalli, 2015, An efficient ant colony system for edge detection in image processing, in: Proceedings of the European Conference on Artificial Life, pp. 398–405.

[29] M. Nayak, P. Dash, Edge detection improvement by ant colony optimization compared to traditional methods on brain MRI image, Commun. Appl. Electron. 5 (8) (2016) 19–23.

[30] C. Pereira, L. Gonçalves, M. Ferreira, Exudate segmentation in fundus images using an ant colony optimization approach, Inf. Sci. (Ny). 296 (2015) 14–24.

[31] C. Ledig, L. Theis, F. Huszár, J. Caballero, A. Cunningham, A. Acosta, W. Shi, Photo-Realistic Single Image Super-Resolution Using a Generative Adversarial Network, arXiv preprint, 2016.

[32] S. Goel, A. Verma, K. Juneja, 2015. A framework for improving misclassification rate of texture segmentation using ICA and ant tree clustering algorithm, in: Computing, Communication & Automation (ICCCA), 2015 International Conference on IEEE, pp. 22–27.

[33] Z. Zhang, Q. Wu, Z. Zhuo, X. Wang, L. Huang, Wavelet transform and texture recognition based on spiking neural network for visual images, Neurocomputing 151 (2015) 985–995.

[34] D. Avci, M. Poyraz, M.K. Leblebicioğlu, An expert system based on discrete wavelet transform-ANFIS for acquisition and recognition of invariant features from texture images, *Signal Processing and Communications Applications Conference (SIU), 2015 23th*, IEEE, 2015, pp. 1070–1073.

[35] E. Berkeley,n.d., Berkeley Segmentation Dataset Images. Retrieved February 23, 2018, from https://www2.eecs.berkeley.edu/Research/Projects/CS/vision/bsds/.

[36] V. Caltech, n.d. Caltech ImageDatasets. Retrieved February 23, 2018, from http://www.vision.caltech.edu/Image_Datasets/Caltech256/.

[37] M. Xu, M. Cong, T. Xie, Y. Tao, X. Zhu, J. Zhao, Unsupervised segmentation of high-resolution remote sensing images based on classical models of the visual receptive field, Geocarto. Int. 30 (9) (2015) 997–1015.

[38] M. Mahi, Ö.K. Baykan, H. Kodaz, A new hybrid method based on particle swarm optimization, ant colony optimization and 3-opt algorithms for traveling salesman problem, Appl. Soft Comput. 30 (2015) 484−490.

[39] Y. Chen, A. An, Application of ant colony algorithm to geochemical anomaly detection, J. Geochem. Explor. 164 (2016) 75−85.

[40] R. Grycuk, M. Gabryel, R. Scherer, S. Voloshynovskiy, Multi-layer architecture for storing visual data based on WCF and microsoft SQL server database, International Conference on Artificial Intelligence and Soft Computing, Springer, Cham, 2015, June, pp. 715−726.

[41] A. Biniaz, A. Ataollah, Segmentation and edge detection based on modified ant colony optimization for Iris image processing, JAISCR 3 (2) (2013) 133−141. Available from: https://doi.org/10.2478/jaiscr-2014-0010.

FURTHER READING

S. Sarkar, S. Das, S.S. Chaudhuri, A multilevel color image thresholding scheme based on minimum cross entropy and differential evolution, Pattern Recogn. Lett. 54 (2015) 27−35.

Chapter 3

A Feature Fusion-Based Discriminant Learning Model for Diagnosis of Neuromuscular Disorders Using Single-Channel Needle Electromyogram Signals

A. Hazarika and M. Bhuyan
Department of Electronics and Communication Engineering, Tezpur University, Tezpur, India

Chapter Outline

3.1 INTRODUCTION

Availability of big data [1] in various processes and its requirement for accurate and reliability assessment tremendously increases research on feature fusion models [2,3]. Nonlinear process assessments involve a large number of measurements or signals, which are a wide variety in nature. Each signal or dataset poses a different degree of uncertainty and does not have the same level of confidence, reliability, and information quality. A single set-based measurement may not ensure the physiological and anatomical behavior of

Intelligent Data Analysis for Biomedical Applications. DOI: https://doi.org/10.1016/B978-0-12-815553-0.00003-3
49

the process. As a result, big data or an enormous volume of data usage by making an appropriate sense becomes the key to derive a reliable and accurate data-driven model [4]. Model failure occurs when it is unable to cope with relevant information from available resources. Thus proactive data processing is necessary in order to enhance the uniqueness, interpretability, and robustness of models [5].

Large-scale measurements contain a wide range of diagnostic information and, therefore, use of evidence, obtained with the aid of a signal-processing approach, enhances the reliability of model outcomes. In Ref. [1], usefulness of big data in various applications, including industry and healthcare, was discussed. A comprehensive review in [6] showed the recent development of big data in the context of biomedical and health informatics. Ding et al. proposed a big data algorithm for risk analysis of industrial processes [7]. Usually the lack of information contents, parameter constraints, and inappropriate feature dimensionality cause failure and outages of models. Decision support systems rely highly on an appropriate feature extraction framework and feature dimensionality. Therefore research aims to provoke interest to develop such design fronts by allowing interaction of multiple measurements obtained from the same or different modalities. The features based on such frameworks could enable an efficient characterization of the process. Solving this challenge presents a unique opportunity for more effective prediction and diagnosis model designs. It requires effective implementation and optimized systems in order to reduce the computational bottleneck of handling these data. Consequently, feature fusion [8−12] has become essential to realize the envisioned efficient model. It provides such a well-defined framework so as to extract suitable low-order features that contain much richer information about the underlying process.

The wide variability and nonstationary nature of biomedical signals are inherent properties. For instance, electromyogram (EMG) and electroencephalogram (EEG) signal frames of the same category subject group in Fig. 3.1 reveal the facts. As a result, features extracted from specific

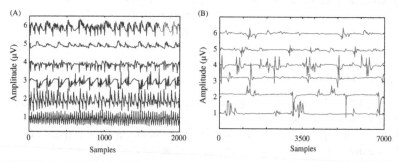

FIGURE 3.1 (A) Multiple EEG signals having seizure activity and (B) EMG signals having ALS disorders. One-dimensional signals are collected from two different sources: (A) from http://www.meb.unibonn.de/epileptologie/science/physik/eegdata.html, (B) from http://www.emglab.net/.

Multiple 1D signal

FIGURE 3.2   Block diagram of proposed learning method using feature fusion strategy.

prediagnosed signals may not accurately represent the underlying disease process. Therefore use of large measurements and embedding suitable features to the data-driven model is essential to avoid feature biasing and to accurately predict various inputs. Such requirements need the to development of computer-aided diagnosis for objective assessments of underlying diseases' process and monitoring of patients' health status.

This chapter presents a discriminant, feature fusion real-time, data-driven model generalizing the discriminant correlation analysis (DCA). Fig. 3.2 shows the block diagram of the proposed approach. The main contributions of this chapter are:

1. A high-dimensional input feature generation scheme is proposed that can employ large-scale available measurements which contribute to the implementation of feature fusion based model.
2. An efficient fusion-based system that integrates high-dimensional features for optimization and to obtain high-level abstraction for prediction and diagnosis of neuromuscular disorders.
3. The adopted algorithm is tested in offline mode using nonlinear data and demonstrated the effectiveness for real-time implementations.
4. The proposed algorithm's performance is compared with the state-of-the-art methods in the context of feature extraction and performance.

The remainer of the chapter is organized as follows. Section 3.2 outlines the state-of-the-art methods. Section 3.3 provides the theoretical modeling of big data learning. Section 3.4 provides a brief overview on medical measurements and data analysis. Section 3.5 provides results and discussion. Finally, Section 3.6 presents the conclusions reached.

3.2   STATE-OF-ART-METHODS

Quantitative decision support models (QDSM) employ quantitative information or features elicited from image [13−19] or one-dimensional signals. Unlike traditional qualitative assessments that require expert domain-specific knowledge, QDSMs provide quantitative information, such as disease severity. As a result, extensive research efforts have been made to provide a more sophisticated diagnosis probe that will help to gain an in-depth understanding of the underlying process of the disease and subsequent diagnosis.

EMG-QDSM employ quantitative features elicited from signals to diagnose various neuromuscular disorders, such as *amyotrophic lateral sclerosis* (ALS)

[20] and myopathy [21]. For instance, Yousefi et al. [22] outlined various methods that show steady improvement in classification performance. Many earlier methods employed motor unit action potential (MUAPs) features [23], wavelet-feature [24,25], autoregressive coefficient (AR), time-domain (TD) [26], AR + RMS [27,28], and multiscale PCA features [29] and intrinsic mode functions [30] extracting from signals with special emphasis on the models. However, due to the typical nature of signals, model constraints, and computational bottlenecks, many methods failed to achieve the goal. In [31], various MUAP feature extraction techniques and performance of soft computing models were discussed, which showed that the adaptive neuro-fuzzy inference system is one of the promising approaches for classification tasks. Subasi [32] adopted a wavelet-based fuzzy support vector machine (SVM) for classification of normal, neurogenic, or myopathic disorders. However, choice of an appropriate wavelet function and coefficients for an effective model often requires human intervention [30]. Many approaches employed a frequency feature using Short-Term Fourier Transformation (STFT). However, due to poor resolution, STFT is not popular for nonstationary signal analysis. Unlike such transformation, wavelet transformation-based methods provide good time-frequency resolution for high and low frequency components of a signal [33]. In Ref. [34], a hybridized approach of particle swarm optimization and SVM was proposed that employed frequency features.

Recently, Doulah et al. [35] adopted two schemes by employing high-energy wavelet coefficients from signal frames and dominated MUAPs respectively, for classification of disorders using k-nearest neighbors (k-nn) and reported promising results. However, the extraction of dominated MUAPs from signals requires guidance information of class-specific patterns. In Ref. [36], an evolutionary approach was proposed through optimized kernel parameter setting using a GA-algorithm and reported an accuracy of 97.0%. Gokgoz et al. [37] introduced a decision tree algorithm that extensively used an in-machine learning community and reported an accuracy of 96.67%. More recently, Hazarika et al. [11,12] researched CCA-fusion models for EMG classification and reported promising performances.

Despite substantial improvements of data-driven techniques in comparison to traditional machine learning approaches, there are still many research challenges. Challenges arise since learning frameworks have not yet provided complete understanding of various processes and solutions and many research queries remain unanswered. In order to design an effective and viable diagnosis model, big data employment for learning is essential which could embed complete process information. Such a proactive strategy is also helpful for personalized mobile healthcare, bioinformatics, and sensor informatics model designs [6]. An appropriate signal processing approach by employing statistical models can make a significant impact in this direction. The most important aspect to account for while developing data-driven models is that for many applications the raw data cannot be directly embedded as input due to its large dimensionality. Furthermore, raw data contains superfluous information which must be removed at the initial stage. Thus prepossessing, transformation, normalization, and optimization must be emphasized and implemented for a viable model derivation.

3.3 THEORETICAL MODELING OF LEARNING FROM BIG DATA

3.3.1 Strategy Statement

The main objective of big data employment is to extract relevant quantitative information or features that can provide complete understanding of underlying phenomena. Usually nonlinear process assessment and diagnosis inherently includes a large number of measurements or signals. Each signal adds information about the underlying process. Our hypothesis is, therefore, to create a set of multi-view features (MVs) by employing multiple signals through a specific strategy which could have the most physiologically relevant information and energy contents. The low-order feature structure obtained from such high-dimensional space using a signal processing technique could have more potential benefits and presumably avoid feature biasing in the learning framework. As is evident, discrete wavelet transformation (DWT) is a profound data processing approach for nonlinear analysis. Unlike previous methods that employed wavelet features directly for classification tasks, we adopted DWT to evolve statistically independent MVs. The idea is to extract statistically optimized features from directly evaluated MVs and its corresponding wavelet MVs. It then, aims to fuse independent features to enhance the quality of features in the context of its generalization ability that leads to obtain optimum model performance [38].

For example, consider a clinical medical dataset that consists of C subject groups and each group includes multiple signals recorded during the same, or different, experimental conditions. Assume that there are wide variety in signals within each group. Therefore each C is partitioned into G subgroups from where corresponding feature matrices $\mathbf{X}$s, also known as MVs, are evaluated. These feature matrices are further decomposed for correlation-based analysis. It is due to the fact that CCA and DCA can effectively find inherent mutual information between two consecutive feature variables while dissimilar components are eliminated. Unlike CCA, DCA contains a feature discriminant function which makes it an attractive choice for classification tasks [8].

3.3.2 Discriminant Feature Fusion Framework

An input MV $\mathbf{X}$ composed of N samples with m signals for a particular input process and $\mathbf{Z}$ is the corresponding MV in wavelet space

$$\mathbf{X} = [x_1, x_2, \ldots, x_N]^T \in \mathbb{R}^{N \times m} \tag{3.1}$$

$$\mathbf{Z} = [z_1, z_2, \ldots, z_N]^T \in \mathbb{R}^{N \times m} \tag{3.2}$$

and $x_n \in \mathbb{R}^m, z_n \in \mathbb{R}^m, n = 1, \ldots, N$. Each z_n in $\mathbf{Z}$ is the low-frequency component of corresponding element in $\mathbf{X}$. Low frequency components y_ns (i.e., A_n) are obtained by performing DWT over each x_n in $\mathbf{X}$. However, high-frequency components D_ns are ditched from analysis which will be explained later. The MVs in Eqs. (3.1)–(3.2) are uniformly decomposed into a set of vectors, termed as decomposed-MV (dMV)

$$\mathbf{X} = [X_1, Y_2, \ldots, X_k]^T, \quad X_k \in \mathbb{R}^{N \times l} \tag{3.3}$$

$$\mathbf{Z} = [Z_1, Z_2, \ldots, Z_k]^T, \quad Z_k \in \mathbb{R}^{N \times l} \tag{3.4}$$

and $l = m/k$, k being number of decomposition. For mathematical simplicity consecutive dMVs X_1, Y_2 are simply indicated by X and Y. In order to remove the redundancy from the inputs PCA [39–41] dimension reduction is applied [8]. Furthermore, the mean of each row from the obtained dMVs are removed to make centered data matrices [42]. CCA project the pair of input dMV into subspace to find a linear transformation [43]

$$\left. \begin{array}{l} u = A_{x_1}x_1 + \cdots + A_{x_k}x_k = A_x^T X \\ v = B_{y_1}y_1 + \cdots + B_{y_k}y_k = B_y^T Y \end{array} \right\} \tag{3.5}$$

The projected vectors, $A_x \in \mathbb{R}^d$ and $B_y \in \mathbb{R}^d$ are evaluated from orthogonal subspaces by optimizing the following correlation criteria

$$\max_{A_x, B_y} \rho(u, v) \rightarrow \frac{A_x^T \Sigma_{xy} B_y}{\sqrt{(A_x^T \Sigma_{xx} A_x)(B_y^T \Sigma_{yy} B_y)}} \tag{3.6}$$

$$s.t. A_x^T \Sigma_{xx} A_x = B_y^T \Sigma_{yy} B_y = 1 \tag{3.7}$$

where Σ_{xx} and Σ_{yy} are autocovariance matrices, and Σ_{xy} and Σ_{yx} are cross-covariance matrices of X and Y. The optimization requires solving two eigenvalue equations

$$XY^T(YY^T)^{-1}YX^T A = \rho^2 XX^T A \tag{3.8}$$

$$YX^T(XX^T)^{-1}XYB = \rho^2 YY^T B \tag{3.9}$$

$$\Sigma_{xx} = XX^T, \quad \Sigma_{yy} = YY^T, \quad \Sigma_{xy} = XY^T, \quad \Sigma_{yx} = YX^T \tag{3.10}$$

However it can be solved using singular value decomposition (SVD) [44] which evaluate the projection vectors by transforming SVD of $X^T Y \in \mathbb{R}^{d \times d}$

$$X^T Y = U \Lambda V^T = [U_1, U_2]D[V_1, V_2]^T = U_1 \Lambda_d V_1^T \tag{3.11}$$

$$d \leq N = \text{rank}(X, Y), X^T U_2 = 0, A_x = XU_1, B_y = YV_1 \qquad (3.12)$$

U and V are two left and right singular orthogonal matrices of Σ_{xy} and Σ_{yx}, respectively. Here $D = \text{diag}(\Lambda_d, 0)$ eigenvalue matrix whose diagonal element representing the correlation of project vectors obtained by Eq. (3.12) are in descending order. These high-order projected vectors correspond to high correlations in D and well-preserves the information from input features. Finally, CCA derive two discriminant features using parallel and serial fusion strategies which results in significant reduction of feature dimensionality.

3.3.3   Generalized Multidomain Learning

The main limitation in CCA is the ignorance of class-structure information which is essential for multitask learning models. The edited version of CCA, namely, DCA [8] employs the class-structure information, which simultaneously correlates within-class features and decorelates between-class features. In other words, it maximizes the between-class feature matrices of transformed features and diagonalizes the individual between-class matrix to ensure well-class separation. Similar to the CCA, DCA canonical variates for discriminant derivations summarizes the correlation structure among variables using the class-structure information. However, it does not include domain independent features while deriving discriminant features using fusion strategies. In order to obtain a more generalized version in contrast to the DCA, an extended version, namely, generalized multidomain multi-view-DCA, is employed. This model is further regulated by appropriately choosing the regularization parameters. It aims to evaluate domain-independent features using criteria functions and employs them to find the discriminant feature generalizing the feature fusion schemes for the learning task [45,46]. It finally derives two sets of transformation $\bar{u} = A_x^T X$, $\bar{v} = B_y^T Y$ using the feature-discriminant function, as used in DCA-based model [8,47]. In contrast, this multi-domain learning framework finds another set $\bar{u}_{z1} = C_{z1}^T Z_1$, $\bar{v}_{z2} = D_{z2}^T Z_2$ for two wavelet-based inputs Z_1 and Z_2. These two sets are fused independently using the summation technique [8] and combines into a single vector T_{ij}.

$$T_{ij} = \sum_{i=1}^{2} T_t(\bar{u}, \bar{v}) = T_1(\bar{u}, \bar{v}) + T_2(u_{z1}, v_{z2}) \qquad (3.13)$$

$$T_1(\bar{u}, \bar{v}) = A_x^T X + B_y^T Y, T_2(u_{z1}, v_{z2}) = C_{z1}^T Z_1 + D_{z2}^T Z_2 \qquad (3.14)$$

where C and D are weight vectors for two wavelet MVs and T_{ij} indicates generalized features (gDF) for each pair of dMV. This way, gDFs are evaluated for all pairs of dMVs based on which more favorable statistics are evaluated.

3.4 MEDICAL MEASUREMENTS AND DATA ANALYSIS

3.4.1 Electromyogram Signal Recording Setup

Fig. 3.3 shows the EMG recording set up. A typical multiple electrode system is used to acquire the electrical responses of muscles belonging to various parts of human body. The electrical responses, known as EMG signals, are used to assess the abnormalities of muscles. However, the analysis involves multiple signals recorded from various regions of muscles under investigation for proper interpretation of disorders based on some key features. Based on measurements and inherent characteristics, the process under investigation is termed abnormal or normal. The abnormal process, or disorder, is further categorized depending on the nature of the abnormalities. Unlike traditional qualitative methods, quantitative methods are dominating nature in providing quantitative insights of the processes.

The recording set-up includes: signal, signal processing unit, signal-output, and data storage system. Recordings are done with a connection of inbuilt active, reference, and ground electrode systems and an external temperature control probe. The signal processing unit includes high-gain differential amplifiers, filters, and an analog-to-digital converter (ADC). It amplifies the potential difference between the active and reference inputs to improve the SNR. The amplifier has high input impedance $> 10^3$ MΩ and a high common-mode rejection ratio, for example, >100 dB. It has the ability to acquire the signal in the range of 1 μV$-$50 mV. The most useful settings of amplification are 50 μV/cm and 200 μV/cm. All amplifiers use a band-pass filter with an adjustable frequency range to attenuate noise. 1$-$2 kHz is the adjustable lower frequency range, while the notch filter setting is 50 Hz/ 60 Hz mainly to attenuate the power line frequency. ADC converts the analog EMG signal into digital waveforms along with displaying and storing waveforms in a digital format. It has an adequate sampling frequency

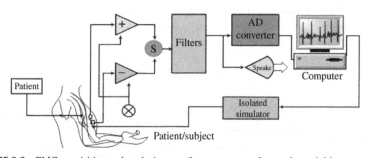

FIGURE 3.3 EMG acquisition and analysis setup for assessment of muscular activities.

capacity to prevent waveform distortion from the aliasing. The signal display unit has sensitivity/gain control, sweep speed adjustment, and adequate vertical/horizontal resolution on the monitor to enable visual assessment of waveforms. The computer with sweep speeds of 510 ms/cm is used to characterize the motor units. An audio amplifier and speaker are connected to the display unit to analyze the characteristic sounds produced during recording and examination [48]. The data management system stores real-time data without subjective details in predefined files for analysis and research.

3.4.2   Electromyogram Datasets

In order to formulate and investigate the performance of the scheme advocated in this study, two clinical online datasets EMG_{S1} [49], and EMG_{S2} collected from Guwahati Neurological Research Centre, Assam, India, are considered. The considered datasets include two disorder subject groups (ALS, myopathy [MYO]) and one healthy control (NOR) subject group. Further it includes male (M) and female (F) participants. Each subject group contains multiple signals recorded during the same experimental setting using needle electrodes with a leading-off area of 0.07 mm^2. The signal was recorded at a sampling rate of 23437.5 Hz with a filter setting of $2-10$ kHz for the duration of 11.2 s, which were amplified with custom built-in-DISA15C01 with a gain of 4000.

The first dataset, EMG_s, includes: 8 ALS (4 M, 4 F, age: $35-67$, mean: 52.8 ± 11.8 (year)), 7 MYO (5 M, 2 F, age: $19-63$, mean: 36.3 ± 14.6) and 10 NOR (6 M, 4 F, age: $21-37$, mean: 27.7 ± 4.5). The second, EMG_{S2}, includes: 4 ALS (1 F, 3 M, age: $38-52$, mean: 43.5 ± 7), 4 MYO (2 F, 2 M, age: $42-59$, mean: 47.5 ± 7.8) and 4 NOR (2 F, 2 M, age: $26-34$, mean: 29.3 ± 3.4). The EMG_{S1} handled 250 signals (150 ALS, 50 MYO and 50 NOR) while the EMG_{S2} handled 60 signals (20 ALS, 20 MYO and 20 NOR) of three subject groups. The frequency of an EMG signal usually falls in the range of $0-1$ kHz, and its dominated energy is confined in the range of $20-500$ Hz [50]. Thereby signals are filtered using a twenty-order Kaiser window-based low pass filter with a cut-off frequency and sampling frequency of 0.5 kHz and 1 kHz [51], respectively. Fig. 3.4 shows three typical signal patterns used in this analysis.

3.5   RESULTS AND DISCUSSION

According to the problem statement in Section 3.3.1, $G = 3$ is considered for each study group depending on the subjects' age and formulated the MV features with randomly selected signals from the respective subgroup. Furthermore, according to the concept of the GMDCA, second-order discreet wavelet transformation with prototype $db2$ [35] is performed over MVs and

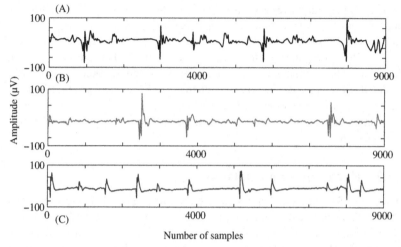

FIGURE 3.4 Three typical signal patterns: (A) ALS, (B) myopathy, and (C) healthy control subject.

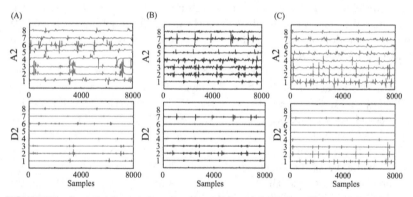

FIGURE 3.5 Second order wavelet-approximate (A2) and detailed coefficients (D2) with prototype *db2*: (A) ALS, (B) myopathy, and (C) normal with limited samples.

taking only low-frequency components MVs and dMVs are evaluated, as stated in Eqs. (3.2) and (3.4). The high-frequency components are ditched from the analysis due to the fact that information contents of most real-time signals fall in the low-frequency scale. Fig. 3.5 shows the wavelet components for three studied inputs. It is worth noting that we employ wavelets only to transform MVs to wavelet-domain and to extract statistically independent MVs. Therefore instead of taking multiple wavelet families, only *db2* is used in this analysis, which is shown to be promising for EMG analysis [35].

3.5.1   Correlation Analysis

Existence correlation between input features ensures the efficacy of correlation-based analysis [8]. The correlation analysis among various consecutive pairs of dMV are performed similar to the methods in Refs. [12,43,52]. Low correlation-based statistics may fail to explicitly represent the inherent information of features. Thus correlations have been demonstrated for an appropriate consideration of low-order statistics for classification tasks. It has been analyzed for different choice of $m = 8, 10, 12$ and three feature sets corresponding three values of m each with four dMVs— that is, $k = 4$ as in Eqs. (3.3)–(3.4)—are evaluated for each group. The dimensions of dMV are $8 \times 64,500$ for $m = 8$, $10 \times 64,500$ for $m = 10$, $12 \times 64,500$ for $m = 12$. However, results are presented for $m = 8$ which is statistically significant ($P < .05$) as investigated during performance analysis. The correlations for consecutive pair dMVs, that is, $(1, 2)$, $(2, 3)$, $(3, 4)$, and $(4, 5)$ of respective processes are estimated, as shown in Fig. 3.6. The dMV for $m = 10, 12$ is the extension of dMV for $m = 8$ since these dMVs are obtained by an incremental increase of m by 2. As is evident, a correlation exists among the group-specific dMVs up to a reasonable level. However, correlations among lower-order transformed features are insignificant. The features corresponding to high-order correlations are considered in this study.

3.5.2   Performance Investigation of Discriminant Learning Scheme

The gDFs are evaluated for each pair of dMVs using Eq. (3.14) in contrast to the standard DCA that employs single-domain features in the same procedure. However, in evaluating the gDFs the statistics for correlation threshold is at 7 in two independent analyses. This way four and twelve number of gDFs are evaluated for each G and C. The obtained group-specific mean gDFs are transformed using a proposed linear transformation $\Theta_i = T_{ij} + \delta I$, δ with I being the class-indicator parameter and unitary matrix. Such a

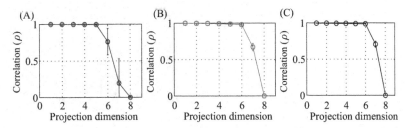

FIGURE 3.6 Mean correlations among the dMVs of three groups with $\pm$ SD as error bars: (A) ALS, (B) myopathy, and (C) healthy control subject or normal.

transformation improves the robustness of feature space and helps retain the quality [53]. The dimension of T_{ij} is matched with I by padding zeros with an empirical setting of $\delta = 0$, 10, and 20 for normal, myopathy, and ALS respectively.

The obtained generalized feature space is subjected to linear discriminant analysis (DA) so as to obtain high marginal features. It employs class-structure information with the use of within-class variance and maximizes between-class variance and finds the linear transformation to set the best decision margin among multigroup features. A one-way analysis of variance (ANOVA) is carried out and P-values of features are evaluated. A set of statistically significant features $f_1 - f_5$ are selected with P-value $< 1 \times 10^{-4}$. However, intercross validation with training data and ANOVA show that the combination $f_1 - f_4$ is capable of providing optimal performance. This analysis further shows that the P-values of features obtained without using the proposed linear transformation are also statistically significant. The analysis has been carried out only at $m = 8$ since statistical testing reveals that there was no benefit in adding more than eight signals while deriving input features. Fig. 3.7A shows the box-plot feature distribution which are embedded to the simple decision models.

The entire dataset $EMG_{S1} + EMG_{S2}$ is partitioned into three subsets—training set (50%), validation set (25%), and test set (25%)—for performance evaluation. The training dataset is used for evaluating the optimal feature combination which is then employed for performance estimation using intercross validation strategy. Thus the model has been supplied with discriminant features to be assigned any one of three input processes: ALS, myopathy, and normal. Choice of simple models to integrate with the proposed scheme is mainly due to two reasons. First, it provides easy accessibility and understanding for practical uses. It is worth mentioning that the classification performance highly depends on data integration and feature extraction strategies rather than the choice of classifiers [5]. Second, it ensures whether the proposed feature extraction technique is capable of

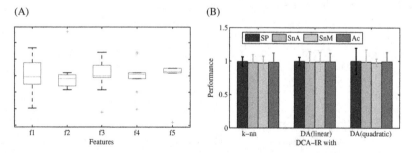

FIGURE 3.7 (A) Box-plot of feature distribution and (B) classification performance of various classifiers in percentage ($\times$ 100).

covering wide variety large-scale information in real-world applications, since specific algorithm performance may not ensure the efficacy of the selected feature space.

k-nn [54] the requires only tuning parameter k, is adopted with choice of $k = 1$. The model performances have been measured in term of conventional biomedical markers-accuracy (Ac), sensitivity (Sn), and specificity (Sp) [55−57] similar to the performance indicators like product quality [58]. In the streamline evaluation process, an additional classifier-DA model with linear and quadratic discriminant functions is also considered and, subsequently, marker parameters have been presented in Fig. 3.7B. As evident, the proposed algorithm achieves promising performance with multiple integrations. The DA scheme with quadratic function achieves slightly better performance, presumably due to the parameter variation of models. The low variance in outcomes indicate small feature biasing in the initial framework of learning models. In other words, the proposed feature extraction scheme employ large volume data that leads to obtain suitable features using the optimization technique. Thus the weight of redundancy on algorithm performance is significantly reduced and obtained low-dimensional statistics well-facilitates the learning task. Another integrity of the adopted scheme is that because of the small dimensionality of feature space it is easy to adopt various decision models. Quantitatively the algorithm achieves an optimal accuracy of 98% with a specificity of 99.30% and sensitivities of 97.61% and 97.14%. However, in assessing the performances two sensitivities—SnA, SnM for two positive cases—ALS and myopathy are evaluated separately [38,52]. The obtained results with low computational time multiple iterations, that is, 15−20 seconds, ensure the suitability of our scheme.

3.5.3   Comparative Study

This section briefly highlights the efficacy of the adopted learning model in the context of some relevant state-of-the-art methods. As discussed in Section 3.2, various approaches employed morphological features, wavelet-statistical features [32,34,37], multiscale PCA [29], intrinsic mode functions [30], autoregressive coefficients, and discriminant CCA features [3]. They have pros and cons when applied in real-time applications. The method in [29] requires optimal kernel function parameters (C and γ) for accurate classification, which is tricky. In Refs. [34] and [29], the authors proposed hybridized PSO-SVM and multiscale-PCA which were derived from evolutionary algorithms. Two characteristic limitations of these algorithms are computational burden and lack of theoretical guarantees. Subasi adopted a decision support system that employed a GA-algorithm to select optimum kernel parameters [36]. However, it involves a number of steps. The decision tree support framework [37] takes large memory and also slows down the process in the case of a large number of trees. Further, lack of explanation and a tendency to overfit while classifying noisy data are other difficulties of this approach.

Besides, many nonstationary signal-based approaches employed frequency features [35] and time-frequency [21] extracting from specific diagnosed signals. These features are promising in carrying the underlying information of signals. However, due to wide variability in data patterns, even within the same subject, they may fail to represent the process associated with signals. Frequency features are obtained by using STFT and wavelet transforms (WT). However, due to poor resolution and expensive computational cost, STFT is not commonly used. In wavelet-based analysis method, an appropriate wavelet function and coefficients for effective model performance are tricky and often requires human intervention [30]. In the case of EMG support systems, it is difficult to match matching wavelet coefficients from each level with the specific MUAP.

Big data learning is gaining attention in the machine-learning community. Large-scale implementations to solve many complicated real-world problems enhances the reliability of the algorithm and diagnosis devices that rely on it. This chapter demonstrated how such a learning approach has enabled the development of more efficient data-driven models for real-world applications, including medical and industrial, by employing a large amount of data for appropriate integration to extract features, that reduce human intervention and the burden of error-prone conventional analysis. Specifically, in the medical field, healthcare devices based on such an efficient diagnostic algorithm provide more reliable and faithful decisions to enhance and streamline business. Most of the approaches focus on evaluating low-dimensional features to avoid the learning constraints; however, due to lack of information in such feature space, they often fail to achieve this goal. Taking, as a key challenge, handling large-volume data, the proposed approach developed an efficient framework and implemented real-time data.

This study, thus, demonstrates how assessment of multiple measurements through an effective data integration framework helps to understand real-world phenomena. Although the typical nature of biomedical signals complicates the process understanding, a data-driven approach with a well-defined initial framework that allows the interaction of heterogeneous measurements to obatin suitable features for model learning provides the key path to characterize the process and ensures efficacy of design. In contrast to other methodologies, the method advocated in this chapter reveals the suitability of high-level abstraction strategy to solve complicated nonlinear problems, which involve mainly processing an enormous volume of data, and optimization and fusion without prior knowledge of the subject's background. It is seen that it outreaches many theoretical and computational bottlenecks and outperforms many state-of-the-art methods.

3.6   CONCLUSION

The chapter presented an efficient data-driven model based on feature fusion framework by using multi-view features obtained from large-volume data. First, a multi-view feature formulation strategy for high-level abstraction of information was proposed. The chapter then advocated a feature fusion-based model for nonlinear medical diagnosis problems. An experiment was carried out using real-time EMG data for offline prediction of algorithm performance. The promising performance over repeated experiments and comparison analyses reveal the effectiveness and feasibility of the discriminant learning framework to accelerate real-world inference systems.

REFERENCES

[1]   Z. Lv, H. Song, P. Basanta-Val, A. Steed, M. Jo, Next-generation big data analytics: State of the art, challenges, and future research topics, IEEE Trans. Ind. Inform. 13 (2017) 1891−1899.

[2]   L. Dana, T. Adali, C. Jutten, Multimodal data fusion: an overview of methods, challenges, and prospects, Proc. IEEE 103 (2015) 1449−1477.

[3]   A. Hazarika, M. Barthakur, M. Bhuyan, Fusion of projected feature for classification of EMG patterns, Proceedings of the IEEE Conference on recent Advances and Innovations in Eng., IEEE, India, 2016, pp. 69−74.

[4]   J. Wan, S. Tang, S. Wang, C. Liu, H. Abbas, A.V. Vasilakos, A manufacturing big data solution for active preventive maintenance, IEEE Trans. Ind. Informat. 13 (2017) 2039−2047.

[5]   J.L. Hargrove, K. Englehart, B. Hudgins, A comparison of surface and intramuscular myoelectric signal classification, IEEE Trans. Biomed. Eng. 54 (2007) 847−853.

[6]   J. Andreu-Perez, C.C. Poon, R.D. Merrified, S.T. Worng, G.Z. Yang, Big data for health, IEEE J. Biomed. Health Inform. 19 (2015) 1193−1208.

[7]   X. Ding, Y. Tian, Y. Yu, A real-time big data gathering algorithm based on indoor wireless sensor networks for risk analysis of industrial operations, IEEE Trans. Ind. Inform. 12 (2016) 1232−1242.

[8]   M. Haghighat, M.A. Mottaleb, W. Alhalabi, Discriminant correlation analysis: Real-time feature level fusion for multimodal biometric recognition, IEEE Trans. Inform. Forensics Secur. 11 (2016) 1984−1996.

[9]   R.C. Luo, C.C. Chang, C.C. Lai, Multisensor fusion and integration: theories, applications, and its perspectives, IEEE Sens. J. 11 (2011) 3122−3138.

[10] R.C. Luo, C.C. Chang, Multisensor fusion and integration: a review on approaches and its applications in mechatronics, IEEE Trans. Ind. Inform. 8 (2012) 49−60.

[11] A. Hazarika, L. Dutta, M. Barthakur, M. Bhuyan, Two-fold feature extraction technique for biomedical signals classificationAugust Proceedings of the IEEE Conference, in Inventive Computation Technologies, IEEE, India, 2016pp. 1−4.

[12] A. Hazarika, L. Dutta, M. Boro, M. Barthakur, M. Bhuya, An automatic feature extraction and fusion model: application to electromyogram signal classification, Int. J. Multimed. Inform. Retr. (2018). Available from: https://doi.org/10.1007/s13735-018-0149-z.

[13] J.D. Hemanth, E.B. Valentina, Special Issue on Decision Support Systems for Medical Applications, Intelligent Decision Technologies, IOP Press, 2016, pp. 1−2.

[14] J.D. Hemanth, E.B. Valentina, J. Anitha, Hybrid neuro-fuzzy approaches for abnormality detection in retinal images, Soft Computing Applications, Springer International Publishing, 2016, pp. 295–305.

[15] J.D. Hemanth, E.B. Valentina, Performance improved hybrid intelligent system for medical image classification, Proc. on Informatics, ACM, New York, 2015, p. 8.

[16] S.U. Maheswari, J.D. Hemanth, Image steganography using Hybrid Edge detector and ridgelet transform, Def. Sci. J. 65 (2015) 214–219.

[17] S.U. Maheswari, J.D. Hemanth, Frequency domain QR code based image steganography using Fresnelet transform, AEU-Int, J. Electron. Commun. 69 (2015) 539–544.

[18] J.D. Hemanth, C.K.S. Vijila, A.I. Selvakumar, J. Anitha, Performance improved iteration-free artificial neural networks for abnormal magnetic resonance brain image classification, Neurocomputing 130 (2014) 98–107.

[19] J.D. Hemanth, C.K.S. Vijila, A.I. Selvakumar, J. Anitha, Distance metric-based time-efficient fuzzy algorithm for abnormal magnetic resonance brain image segmentation, Neural Comput. Appl. 22 (2013) 1013–1022.

[20] A.J. Waclawik, Neurodegenerative disorders: amyotrophic lateral sclerosis and inclusion body myositis, Neurol. Board Rev. Man. Neurol. 8 (2004).

[21] T. Kamali, R. Boostani, H. Parsaei, A multi-classifier approach to MUAP classification for diagnosis of neuromuscular disorders, IEEE Trans. Neural Syst. Rehabil. Eng. 22 (2014) 192–200.

[22] J. Yousefi, A.H. Wright, Characterizing EMG data using machine-learning tools, Comput. Biol. Med. 51 (2014) 1–13.

[23] C.D. Katsis, T.P. Exarchos, C. Papaloukas, Y. Goletsis, D.I. Fotiadis, I. Sarmas, A two-stage method for MUAP classification based on EMG decomposition, Comput. Biol. Med. 31 (2007) 1232–1240.

[24] K. Englehart, B. Hudgins, A. Philip, A wavelet-based continuous classification scheme for multifunction myoelectric control, IEEE Trans. Biomed. Eng. 48 (2011) 302–311.

[25] A.P. Dobrowolski, M. Wierzbowski, K. Tomczykiewicz, Multiresolution MUAPs decomposition and SVM-based analysis in the classification of neuromuscular disorders, Comput. Method Prog. Biomed. 107 (2012) 393–403.

[26] K. Englehart, B. Hudgins, A robust, real-time control scheme for multifunction myoelectric control, IEEE Trans. Biomed. Eng. 50 (2003) 848–854.

[27] Y. Huang, B. Kevin, H. Bernard, A.D.C. Chan, A Gaussian mixture model based classification scheme for myoelectric control of powered upper limb prostheses, IEEE Trans. Biomed. Eng. 52 (2005) 1801–1811.

[28] A.D.C. Chan, K. Englehart, Continuous classification of myoelectric signals for powered prosthesis using Gaussian mixture models, Int. Conference on Medicine and Biology Society International Conference, IEEE, Cancun, Mexico, Sep. 2003.

[29] E. Gokgoz, A. Subasi, Effect of multiscale PCA de-noising on EMG signal classification for diagnosis of neuromuscular disorders, J. Med. Syst. 38 (2014) 1–10.

[30] G. Naik, S. Selvan, H. Nguyen, Single-channel EMG classification with ensemble-empirical-mode-decomposition-based ICA for diagnosing neuromuscular disorders, IEEE Trans. Neural Syst. Rehabil. Eng. 24 (2016) 734–743.

[31] A. Subasi, Classification of EMG signals using combined features and soft computing techniques, Appl. Soft. Comput. 12 (2012) 2188–2198.

[32] A. Subasi, Medical decision support system for diagnosis of neuromuscular disorders using DWT and fuzzy support vector machines, Comput. Biol. Med. 42 (2012) 806–815.

[33] Y. Gritli, et al., Advanced diagnosis of electrical faults in woundrotor induction machines, IEEE Trans. Ind. Electron. 60 (2013) 4012–4024.

[34] A. Subasi, Classification of EMG signals using PSO optimized SVM for diagnosis of neuromuscular disorders, Comput. Biol. Med. 43 (2013) 576–586.

[35] A.S.U. Doulah, S.A. Fattah, W.P. Zhu, M.O. Ahmad, Wavelet domain feature extraction scheme based on dominant motor unit action potential of EMG signal for neuromuscular disease classification, IEEE Trans. Biomed. Circuit. Syst. 8 (2014) 155–164.

[36] A. Subasi, A decision support system for diagnosis of neuromuscular disorders using evolutionary support vector machines, Signal Image Video Process 9 (2015) 399–408.

[37] E. Gokgoz, A. Subasi, Comparison of decision tree algorithms for EMG signal classification using DWT, Biomed. Signal Process. Control 18 (2015) 138–144.

[38] A. Hazarika, M. Bhuyan, A twofold subspace learning-based feature fusion strategy for classification of EMG and EMG spectrogram images, in: J. Hemanth, V. Emilia (Eds.), Biologically Rationalized Computing Techniques For Image Processing Applications, Springer, Cham, 2018, pp. 57–84. Chapter 4.

[39] L. Dutta, C. Talukdar, A. Hazarika, M. Bhuyan, A novel low cost hand-held tea flavor estimation system, IEEE Trans. Ind. Electron. (2017).

[40] L. Dutta, A. Hazarika, M. Bhuyan, Direct interfacing circuit-based e-nose for gas classification and its uncertainty estimation, IET Circuits Dev. Syst. 12 (2017) 63–72.

[41] L. Dutta, A. Hazarika, M. Bhuyan, Comparison of direct interfacing and ADC based system for gas identification using E-Nose, Proceedings of the IEEE Conference on Accessibility to Digital World, IEEE, India, 2016.

[42] G. Wang, S. Yin, O. Kaynak, An LWPR-based data-driven fault detection approach for nonlinear process monitoring, IEEE Trans. Ind. Inform. 10 (2014) 2016–2023.

[43] A. Sharma, A. Hazarika, P. Kalita, B.K. Dev Choudhury, A subspace projection based feature fusion: an application to EEG clustering, Proceedings of the IEEE Conference on Signal Proces. and Commun., IEEE, India, January 2017.

[44] L. Sun, S. Ji, J. Ye, Canonical correlation analysis for multilabel classification: a least-squares formulation, extensions, and analysis, IEEE Trans. Pattern Anal. Mach. Intell. 33 (2011) 194–200.

[45] Y.H. Yuan, et al., A novel multiset integrated canonical correlation analysis framework and its application in feature fusion, Pattern Recog. 44 (5) (2011) 1031–1040.

[46] Q.S. Sun, et al., A new method of feature fusion and its application in image recognition, Pattern Recog. 38 (12) (2005) 2437–2448.

[47] A. Hazarika, A. Sharma, A. Boro, B.K. Dev Choudhury, Discriminant correlation based information fusion for real-time biomedical signal clustering, Lecture Notes in Electrical Engineering, Springer, Jan. 2017.

[48] Aanem, Electrodiagnostic study instrument design requirements, (2015).

[49] M. Nikolic, Detailed Analysis of Clinical Electromyography Signals: EMG Decomposition, Findings and Firing Pattern Analysis in Controls and Patients With Myopathy and Amytrophic Lateral Sclerosis (Ph.D. thesis), Univ. Copenhagen, København, Denmark, 2001.

[50] J.U. Chu, I. Moon, J. Lee, S.K. Kim, M.S. Mun, A supervised feature-projection-based real-time EMG pattern recognition for multifunction myoelectric hand control, IEEE/ASME Trans. Mechatron. 12 (2017) 282–290.

[51] R. Arya, S. Jaiswal, Design of low pass FIR filters using Kaiser window function with variable parameter beta, Int. J. Multidiscip. Curr. Res. 3 (2015).

[52] A. Hazarika, M. Bhuyan, M. Barthakur, L. Dutta, Multi-view learning for classification of EMG template, Proceedings of the IEEE Conference on Signal Process. and Commun., IEEE, India, January 2017.

[53] A. Kusiak, Feature transformation methods in data mining, IEEE Trans. Electron. Packag. Manuf. 24 (2001) 214–221.

[54] G.A. Susto, A. Schirru, S. Pampuri, S. McLoone, A. Beghi, Machine learning for predictive maintenance: a multiple classifier approach, IEEE Trans. Ind. Inform. 11 (2015) 812–820.

[55] M. Barthakur, A. Hazarika, M. Bhuyan, A novel technique of neuropathy detection and classification by using artificial neural network (ANN), Proceedings of the IEEE Conference on Adv. Signal Process Commun., IEEE, India, 2013, pp. 706–713.

[56] M. Barthakur, A. Hazarika, M. Bhuyan, Rule based fuzzy approach for peripheral motor neuropathy (PMN) diagnosis based on NCS data, Proceedings of the IEEE Conference on Recent Advances and Innovations in Eng., IEEE, India, May 1–9 2014.

[57] M. Barthakur, A. Hazarika, M. Bhuyan, Classification of peripheral neuropathy by using ANN based nerve conduction study (NCS) protocol, Int. J. Commun. 5 (2014) 31.

[58] X. Xie, W. Sun, K.C. Cheung, An advanced PLS approach for key performance indicator-related prediction and diagnosis in case of outliers, IEEE Trans. Ind. Electron. 63 (2016) 2587–2594.

FURTHER READING

L. Dutta, A. Hazarika, M. Bhuyan, Microcontroller based e-nose for gas classification without using ADC, Sens. Transducers 202 (2016) 38–45.

Chapter 4

Evolution of Consciousness Systems With Bacterial Behaviour

Daniela López De Luise[1], Lucas Rancez[1], Marcos Maciel[2], Bernardo De Elía[3] and Juan Pablo Menditto[3]
[1]CIS Labs, Buenos Aires, Argentina, [2]CAETI, Buenos Aires, Argentina, [3]Escuela Superior Técnica del Ejército, Buenos Aires, Argentina

Chapter Outline

4.1 INTRODUCTION

Ever since people involved robots in human life [1,2] to do repetitive tasks which were previously executed by operators, more specific responsibilities have been developed. For example, saving lives in disaster zones characterized by extreme situations.

In these situations, the structures created by people or nature, are partially or completely destroyed by earthquakes, landslides, cyclones, floods, fires, etc. To overcome these obstacles, innumerable jobs were implemented to grasp these new challenges.

Robots are used to assist and cooperate with people in many applications and projects. A list of examples, ordered in increasing complexity, is listed below:

Intelligent Data Analysis for Biomedical Applications. DOI: https://doi.org/10.1016/B978-0-12-815553-0.00004-5

- Assist people in their daily routines [1,2], e.g.: medication administration, event notification for a doctor's appointment.
- Grocery online shopping according to a list on a smart fridge.
- Autonomous navigation by corridors between plants to replace repetitive agricultural tasks [3]: "The robot starts from an initial point, makes its way between plants until it finishes its goals."
- Environment survey to provide security [4]: "This robot goes through a defined path and detects objects' movements to alert about possible threats."
- Proposed in [5] there is an autonomous robot which detects previously set-up objects in real-time, for instance, a white line on the street goes through a path until it finds a stationary object blocking a pedestrian path. Such examples have no autonomous learning or risky tasks.
- Overcome its own mechanic limitation [6]: "authors created a robot which can climb over a bigger object than the wheel radius. An arm is used to assist itself, useful when there is no alternative path to go forward."

There has been much research focused on improving robots' technical skills. Some proposals are accomplished by mathematical and geometric algorithms, such as recognizing objects through environment semantic segmentation where these objects can be physically represented. It is possible to create a classification of the objects and then quickly take a decision about the path to take because the robot has a higher level of certainty about the environment [7]. As an alternative, detecting and reconstructing a 3D environment using an ICP algorithm based on images' edges characteristics is proposed [8]. On the other hand, with a camera assembled on the robot to record the area characteristics and working out the virtual barycenter, it is possible to determine the direction in which a robot is moving [9,10]. Thanks to the barycenter update the robot can recalculate its position concerning this new barycenter.

There are researchers who classify objects as obstacle or nonobstacle depending on its color [11]. Another method of working with environmental information is dividing it (based on laser) in inspection and direction through an algorithm. Then the information is rebuilt from each recorded frame taking into account the shape of the robot to work out the angle area.

When mathematical algorithms are applied over collected information from devices, skills are provided to robots to overcome limitations and to handle situations that take place in risky environments.

These robots need to complete the following tasks to be useful within a dangerous environment: read volumetric information of the area, read ambient temperature and gas concentration, image analysis and processing, detect danger around the area, send information to an operator, make contact with survivors, etc.

To improve performance, some strategies are listed below:

- A thermal sensor is attached to the robot that detects human body temperature in a Radial Basis Function Network trained in the Dynamic Decay Adjustment model. When a risky situation for the subject is detected, it identifies and isolates the survivor [12].
- Because the robot may be able to detect human body heat, a mounted thermal camera is used to detect human body temperature based on a Radial Basis Function Network trained with the Dynamic Decay Adjustment model to identify and pinpoint survivor locations.
- Another approach related to networks is to determine master nodes conforming a set of communication cells to improve the operator's communication [13].
- An effective method is the use of a data vector conformed by different data gathered by many different sensors in a single robot such as virtual reality (VR), intelligent detection, telecommunications, virtual simulation, environment dynamic modeling, ultrasonic sensor, laser radar, etc., managed and guided by a human operator. As a case study, robot SALVOR [14] is well-suited to protect sensors from harsh conditions and weather (dust, water, or high temperatures).
- Some robots are outfitted with wheels and arms [15] that are not only sensors since they were conceived to assist people and to communicate with industrial, scientific, and medical (ISM) radio bands.
- A specific case is a fire-oriented robot [16] that uses different nodes to establish a proper wireless communication with MANET.

Among other complex tasks, robots dynamically recognize the environment, build a world-map in real-time and move avoiding obstacles. The last activity is definitely the most difficult. There are many relevant advances in this field, some of them detecting concavities in the obstacles [17], merging sensor information to generate better short/long distance world-maps [18], etc.

The millimeter wave radar allows the collection of information even when weather conditions are difficult, which makes it possible to build an updated world-map [19]. Combining different sensor data and algorithms, such as EKF, they can produce recommendations [20]. These EKF and SLAM approaches are used to auto-localize and plan the shortest path to the objective [21].

An additional example is the auto-localization using minimal aggregated sensor errors [22] by implementing a probabilistic localization based on the metric-topology cartography. The multithreading processing [23] allows relating information from many sources.

There are robots with important features concerning equipment and volume, which allows them to operate in very harsh conditions, being self-adaptable and self-reconfigurable, all with an advanced processing capability [24,25].

The robot prototype Movement of Autonomous Robotics Codelet System (MARCOS) implements the learning concept based on technology consciousness via codelets. Each stores environmental information and allows the COFRAM interface to issue recommendations and path plans to avoid objects and achieve the goal. The Robot Task Adviser (RTA) takes action for long-term decisions. These decisions are generated with data collected from the robot sensors, which are processed and subsequently sent to the robot.

The interface has a built-in set of strategies and dynamically created others on consciousness learning. By combining the Bacteria Theory with the RTA consciousness process, a better performance for the codelet is expected, making a more flexible and adaptive model when facing dynamic environments such as disasters zones.

The autonomous mobile Robot MARCOS is based on the consciousness model. It uses a kernel named COdelet FRAMework (COFRAM) to implement the concept of learning (set on consciousness technology) to propose recommendations for guiding the robot's movements [26] in disaster zones.

It is proposed in this chapter to compare the Consciousness Model implemented by MARCOS and a new approach named Bacterial Infection. The test outcome will determine that it is possible to evolve the present consciousness model through originality, new strategies to be considered by the kernel, and better path-planning derived from a more precise, real-time, world-map building.

The rest of this chapter is organized as follows:

- Section 2: Exposition of the working hypothesis, working thesis, and Consciousness theory as well as the features of the MARCOS robot.
- Section 3: Details of the experimental results are exhibited.
- Section 4: Final considerations and future work are presented.

4.2 PROPOSAL

Appearance variability, unpredictability, environmental variations, and sensing noise or unwanted signals are some examples of the wide variety of issues faced by robotics in disaster zones.

Approaches like the ones proposed in [27] or in [28] can work in a close and predefined environment, as they rely on predefined concepts and structured portions of code, but they are not suited for the highly dynamic situations that a disaster environment presents. The prototype MARCOS uses COFRAM as its kernel which implements concept learning based on technology to process the robot's movements.

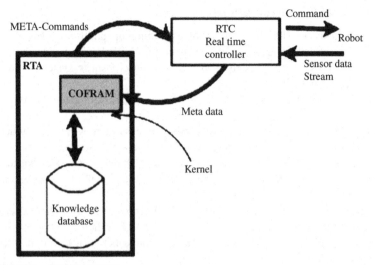

FIGURE 4.1   COFRAM architecture.

The architecture is mainly conformed by a Real Time Controller (RTC) for immediate decisions and a RTA for middle- to long-term strategies [12] as illustrated in Fig. 4.1.

Next is the description of a working hypothesis, working thesis, Consciousness Theory, and finally Bacteria Theory adapted to the present context.

4.2.1   Working Assumptions?

Hypothesis 1. Bacteria behavior generates original alternatives to solve movement strategies in the system context based on consciousness.

Hypothesis 2. Bacteria behavior generates robust alternatives to solve movement strategies in the system context based on consciousness.

Hypothesis 3. The movement strategies' generation process conducted by consciousness is appropriate in real time.

Hypothesis 4. The consciousness model can build an acceptable world map in real time without the need of a previously loaded one.

Hypothesis 5. The consciousness model is flexible to self-adapt to environmental changes automatically.

Hypothesis 6. The autonomous movement model can consider two dimensions to reasonable work in general terms.

Hypothesis 7. The robot autonomy must be such that it allows independent management for dynamic situations in disaster zones, i.e., an abrupt level difference in the path.

Hypothesis 8. The robot is able to record the path, through distance and angle in movement to work out its localization at any given time.

4.2.2 Real Life Assumptions

Hypothesis 1. The infection involves adding up additional behavior to codelets or setting up a predefined function. The behavior is dual:
1. Given by common actions based on consciousness systems (implemented with codelets).
2. Given by the bacteria model creativity which alters the (a) type of actions.

Hypothesis 3. The strategies generation process of movement conducted by consciousness is appropriate in real time, as demonstrated in [12].

Hypothesis 4. It is also demonstrated in [12] that the model based on consciousness builds a world map in real time and can successfully avoid objects.

Hypothesis 5. The consciousness model is flexible to self-adapt to environmental changes automatically. Sanal et al. [12] show the traces where it is possible to verify that the robot can move between different traces. Additionally, the robot achieves the goal.

Hypothesis 7. The robot autonomy must be such that it allows the independent management of dynamic situations in disaster zones, like an abrupt level difference in the path. The autonomy conditions [4,16,29], ability to avoid (Hypothesis 4), and change autonomously strategy [12].

Hypothesis 8. The robot is able to record the path, through distance and angle in movement, to work out its localization at a time x.

It's a standard [12].

4.2.3 Consciousness Theory

Fig. 4.2 shows the consciousness dynamics and Fig. 4.3 the design for its implementation.

According to this, a set of simple small parts of code (named codelets) are probabilistically fired by a manager named coderack. When active, they enter the workspace to reinforce or construct concepts, and/or establish a predefined relationship between them. When a concept (denoted as a "node" in Fig. 4.2) is recalled (activated) many times, it becomes a long-term concept and migrates to the slipnet.

4.2.4 Development of the Bacteria Theory

This section introduces the basis of how the traditional Consciousness Systems adds bacteria adapted to the robot context guided by consciousness elements (codelets).

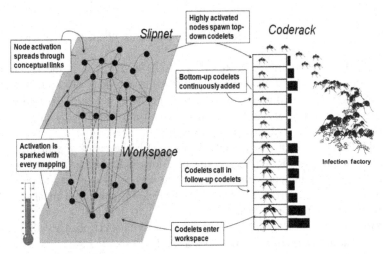

FIGURE 4.2   Consciousness theory.

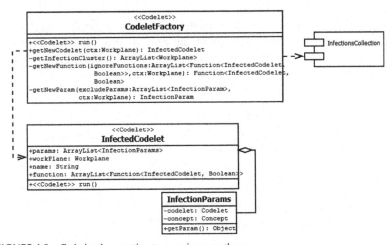

FIGURE 4.3   Code implementation to consciousness theory.

The fundamentals of the CSB theory are the following five steps:

4.2.4.1   STEP 0: Components Set Up

All components in CSB are vectors that evolve as bacteria expands. Vectors must represent the key functional processes in the problem, its parameters, and problem constraints:

$$VS(\text{viral system}) = (VIRUS_i, ORGANISM_i, PROCESS_i)$$
$$VIRUS_i = (STATE_i, INPUT_i, OUTPUT_i, PROCESS_i)$$
$$ORGANISM_i = (STATE_i, PROCESS_i)$$

With:

$STATE_i = (PARAM_1, PARAM_2, \ldots)$ are virus characteristics, actually codelet parameters.

$INPUT_i = (CONCEPT_1, PERCEPTS_2, CODELET_3, \ldots)$ is the organism information that a virus compiles in order to keep track of key functional processes. In the conscious context it is the set of active/passive codelets that perform any activity in the system, and any percept (that is, a concept that comes from environment sensing).

$OUTPUT_i = (CODELET_1, CODELET_2, \ldots)$ is the result of virus activity (represented by $PROCESS_i$). Typically, for consciousness a new set of codelets with unprecedented activity. The new set is expected to produce a change in the system's behavior. In the current context, it means a different ability to learn concepts.

$PROCESS_i = (FUN_1, FUN_2, \ldots)$ virus activity that changes the consciousness functioning, leading to a $STATE_i$.

$STATE_0 =$ organism state at initial moment; it also declares the current status (named clinical profile) and prospective evolution (also referred to as minimum health required).

$PROCESS_0 =$ current conscious model. In biological terminology body-defense, related to the freedom of antigens to protect itself from the infection threat. The application here refers to the capability to change the codelets' abilities.

4.2.4.2 STEP 1: Vectors Initialization

Every vector is assigned a value according to the outlines described in the previous step.

1. Set $INPUT_i = (CONCEPT_1, PERCEPTS_2, CODELET_3, \ldots)$:

   INPUT_i = [
   getHandler(Codelet_1), // CODELETs
   getHandler(Codelet_2),

   ...

   getHandler(Codelet_M),
   getHandler(Percept_actual_1), //PERCEPTs
   getHandler(Percept_actual_2),

   ...

   getHandler(Percept_actual_n),
   getHandler(ConceptName_actual_1 (Class, argv_1...argv_k)), // CONCEPTs
   getHandler(ConceptName_actual_2 (Class, argv_1...argv_k)),

   ...

   getHandler(ConceptName_actual_m(Class, argv_1...argv_k)),
   ]

2. Set $STATE_i = (PARAM_1, PARAM_2, ...)$, i.e., declare codelets' parameters to change their behavior.

$STATE_i = [$

    getHandler(Val_sensor_ultrasound₁),

    getHandler(Val_sensor_ultrasound₂),

    getHandler(Val_sensor_RPLIDAR₁),

    getHandler(Val_sensor_accelerometer₁),

    getHandler(Val_sensor_accelerometer₂),

    getHandler(Val_sensor_accelerometer₃),

    getHandler(Percept_actual₁),

    getHandler(Percept_actual₂),

    getHandler(Percept_actual₃),

    getHandler(Percept_actual₄),

    ...

    getHandler(Percept_actualₙ),

    getHandler(ConceptName_actual₁ (Class, argv₁...argvₖ)),

    getHandler(ConceptName_actual₂ (Class, argv₁...argvₖ)),

    ...

    getHandler(ConceptName_actualₘ (Class, argv₁...argvₖ)),

    ]

3. Set $PROCESS_i = (FUN_1, FUN_2, ...)$

    FUNCIONS = {

    getHandler(FUN₁), // modify sensors values in percepts

    getHandler(FUN₂), // modify concept arguments total

    getHandler(FUN₃), // associate two exists codelets

    getHandler(FUN₄), // duplicate codelets

    getHandler(FUN₅), // change using percept y concepts, using the same function

    getHandler(FUN₆), // concept that gets others concepts and lose any fragment from it-self or build a more simple concept (lose variables)

    }

In this context, the process represents the change of the current codelets set by other codelets (the new ones produced by the infection activity). Several functions for new codelets were considered, many aimed at understanding the shape of eventual obstacles:

FUN(cell) : adjacent? -> pair

FUN(adjacent): continuity? -> collection of adjacent continuous pairs

FUN(continuity): linearity? -> test 4 direct address $+x, +y +x,y x, +y +x,-y$ sabe first and last point

FUN(linearity): curves? -> test lines' adjacent (sharing the first or last point): curves as couples line$_i$, line$_j$

FUN(curves): concave? -> test angle < 180 named CCV

FUN(CCV): concave_continuity? -> common extremes from two or more CCV named concave_curve
FUN (curves):convex? -> test angle > 180 named CVX
FUN(CVX): continuity_convexa? -> common extremes from two or more CVX named convex_curve

It is important to note that considering the out-shape of the obstacle helps to define better strategies for avoiding it.

Some other functions relate to strategies. Since functions are also responsible for creating optimized displacement from one point to another, this is very important to perform efficient movements while avoiding obstacles:

FUN(mov): bring_near? -> verify if it's near to a distance and performs a movement
FUN(mov1, mov2): adjacent?-> sequence of movement SEQ
FUN(SEQ): movement_simplification -> summary the duplicated SEQ in a SEQ-Max

4.2.4.3   STEP 2: Colony Building

The solution set can be characterized by:

$$X = (g\ (\text{INPUT}_k), \dots) \tag{4.1}$$

In Eq. (4.1), $k \in K$, and K is a natural number, X represents the vector of new and potentially more powerful codelets obtained by the application of function $g(x)$ that implements certain heuristics based on the current set of FUNCTIONS that conforms the PROCESS vector.

The parameter k is the index of the definite function with parameters $p_1 \dots p_m$, and $g(x) < = 0$.

Although $g(x)$ could implement any heuristic in order to keep simulating bacteria, it represents the most probable behavior of the virus under current temperature conditions. In this context, it can be the selectability of certain function (Binomial probability of firing a FUN_j) given the degree of conscious activity (**T** parameter):

$$G(x) = \text{Bin}(FUN_j) \times T \tag{4.2}$$

Eq. (4.2) represents the chance of certain set $\{FUN_j\}$ being elected at **T** temperature given the current INPUT (that is a subset of CONCEPTs, PERCEPTs, and CODELETs)

4.2.4.4   STEP 3: Infection Type Definition and Infection

In biological bacteria, the infection produced can be classified as:

1. Selective
2. Massive

To extrapolate this concept to consciousness in robots, let us define $\mathbf{Y}$ as the vector $\mathbf{X}$ size (the number of elements that can be selected for being part of the solution, in other words, the total number of CONCEPTs, PERCEPTs, and CODELETs that constitute the conscious. Let us also define $\mathbf{A}$ as the infections (the number of elements that are affected by bacteria processing; in biology, the number of infected cells).

With the following conditions:

$$A \leq Y$$
$$x \in K$$ (4.3)

The infection type helps to improve flexibility and provide diversity in OUTPUT when it is massive. Conversely, a selective infection helps stabilize results. A combination of both can be implemented using a simple random number:

IF Rnd(1) > 0.5 THEN
   select option (a)
   otherwise select option (b)

4.2.4.4.1 Selective Infection

A single cell is chosen from the entire colony according to selection criteria. In the current proposal a uniform distribution $U(x)$ has been considered.

4.2.4.4.2 Massive Infection

It affects a subset of cells. Let's say $\mathbf{Y\text{-}A}$ cells are infected. The procedure takes $\mathbf{Y\text{-}A}$ elements from INPUT by chance. Infected elements are replaced: $\text{PROCESS}_u\ (x_h)$

In any case, the result of the infection is an OUTPUT vector that must be tested by the conscious system using its Work Plane to evaluate its goodness.

Then the process continues setting $OUTPUT_i = (CODELET_1, CODELET_2, ...)$, the set of new codelets are more abstract. The algorithm for this is:

$OUTPUT_i = [$
   getHandler(Codelet₁), // CODELETs
   getHandler(Codelet₂),
   ...
   getHandler(Codeletₘ),
   ]

Here, $CODELET_i$ is a CODELET that changes its parameters to receive a more abstract function. It is a kind of factory for CODELETs:

```
getNewCodelet(newName){
    NumberOfParameters = getNumberOfParameters(); // returns a total
    of parameters
    parameters = newArray[NumberOfParameters];
    new Class.newInstance(newName, parameters);
}
```

4.2.4.5   STEP 4: Virus Evolution Type Definition

At this time, the biological process continues with cell antagonistic response. The virus evolves in one of the following ways: lysogenic or lytic. To emulate this, "affected" cells (elements) are evaluated as follows:

if Bernoulli (A) > 0 THEN
    The response is lysogenic
otherwise
        The response is lytic

4.2.4.5.1   Lysogenic Replication

When a virus penetrates the host cells and stays in them with no production of new viral particles.

4.2.4.5.2   Lytic Replication

When the virus disseminates, it penetrates the host cell and uses its replication mechanism to produce new viral particles. It starts only if a specific number of nucleus-capsids have been replicated. The number of that the threshold (limit number of nucleus-capsids replication **LNR**) in a cell **x** is given by:

$$LNR_{cell-x} = LNR^0 \left(\frac{f(x) - f(\hat{x})}{f(\hat{x})} \right) \tag{4.4}$$

With:

$f(x)$ = scoring function of the Work Plane (**WP**). Its algorithm is composed by four scoring components:

1. Prioritize forward *straight* line moves when there are no obstacles:
       getScoreNoObstaclesStraightForward(int count, double coeff)
2. where count is the number of obstacles in a straight line, and coeff is defined by:
       coeff = wpStrategy.getEvaluationCoefficients().get(0);
3. Prioritize minimal *distance* until objective is reached:
       getScoreWithObstacles(double distance, double coeff)
       coeff = wpStrategy.getEvaluationCoefficients().get(1);
4. Prioritize *rotations* with forward movements:
       getScoreTurnForward(double angle, double coeff)
       where angle is the minimum angle with no obstacles, and coeff is:
       coeff = wpStrategy.getEvaluationCoefficients().get(2);

5. Prioritize rotations with *backward* movements.

   getScoreTurnBackward(double angle, double coeff)

   where angle is the minimum angle with no obstacles, and coeff is:

   coeff = wpStrategy.getEvaluationCoefficients().get(3);

To perform the evaluation, every WP is checked to evaluate whether a strategy associated to it exists: if this is the case, count in how many **WPs** there is at least one of the codelets listed by **x**. If there is no related strategy then select any **WP** by chance (since at least one of the 0. . . count(**WP**) must be selected).

The cell that produces the better result to the problem is x (in terms of $f(x)$). Here, **x** is the infected cell under consideration and $\mathbf{LNR}^0$ is the initial value of **LNR**.

The infection evolves by cycling all the steps making the systems' elements change progressively. In each iteration there is a number (**NR**) of bacteria replica. The replication number is calculated adding a binomial variable (**z**) to the total value **NR**.

The probability of replicating exactly **z** nucleus-capsids, **P(z)**, the average **E(z)**, and the variance **var(z)** are given by Eqs. (4.5−4.7).

$$p(Z = z) = \left(\begin{array}{c} \mathrm{LNR} \\ z \end{array} \right) p_r^z (1 - p_r)^{\mathrm{LNR}-z}, \qquad (4.5)$$

$$E(Z) = p_r \mathrm{LNR}, \qquad (4.6)$$

$$\mathrm{Var}(Z) = p_r (1 - p_r) \mathrm{LNR}, \qquad (4.7)$$

In this context, p_r is the probability after of a replication and **z** is a random binomial variable, that is:

$$Z = \mathrm{Bin} \ (\mathrm{LNR}, \ p_r) \qquad (4.8)$$

When the edge of a bacteria is broken and releases a virus, each of these released ones have a probability (**p_i**) of infecting neighboring others.

The probability of infecting exactly **y** nucleus-capsids (P(Y = y)), the average (E(Y)), and the variance (Var(Y)) are given by Eqs. (4.9−4.11).

$$p(Y = y) = \left(\begin{array}{c} |V(x)| \\ y \end{array} \right) p_i^y (1 - p_i)^{|V(x)|-y}, \qquad (4.9)$$

$$E(Y) = p_i |V(x)|, \qquad (4.10)$$

$$\mathrm{Var}(Y) = p_i (1 - p_i) |V(x)|, \qquad (4.11)$$

with **IV(x)I** being the feasible solution for the current "environment" or status. In conscious systems it is represented by a set of concepts produced by the sequence of codelets fired previously. **Y** is a random binomial variable which represents the infected cells by the virus in this environment, that is, **Y** = Bin (I$V(x)$I, p_i).

Each of these infected cells in the clinical picture have a probability of developing antibodies against the infection based on the Bernoulli distribution probability (p_{an}: $A(x) = (p_{an})$).

Thus the total infected cells population which produces antibodies are characterized by a binomial distribution given by Eq. (4.12):

$$p_{an}: A(\text{population}) = \text{Bin}\,(n \cdot p_{an}) \tag{4.12}$$

The probability of finding exactly **a** immune cells ($P(A = a)$), the average ($E(A)$), and the variance ($\text{Var}(A)$) are given by Eqs. (4.13–4.15).

$$P(A = a) = \left(\binom{|V(x)|}{a} \right) p_{an}^{a} (1 - p_{an})^{|V(x)| - a}, \tag{4.13}$$

$$E(A) = p_{an} |V(x)|, \tag{4.14}$$

$$\text{Var}(A) = p_{an}(1 - p_{an}) |V(x)|, \tag{4.15}$$

where **a** indicates the immune cell and **x** the infected cell. In the case of a lysogenic type replication, it is possible to calculate the maximum interaction limit (LIT) of a **x** cell with Eq. (4.16).

$$\text{LIT}_{\text{cell}-x} = \text{LIT}^{0} \left(\frac{f(x) - f(\hat{x})}{f(\hat{x})} \right), \tag{4.16}$$

with **LIT$_0$** being the initial value of **LIT**.

4.2.4.6 STEP 5: Closing Step

The algorithm finishes off according to one of these criteria:

A—Collapse or death of the organism (in this case, all the options were tested and there is no more exploration to be done).
B—Virus isolation (in this case the bacteria mechanism is canceled).

4.2.5 Hardware Considerations

The current proposal for MARCOS's hardware is:
1 x Ultrasonic Module-HC-SR04 (Fig. 4.4)
Ultrasonic ranging module HC-SR04 provides 2−400 cm noncontact
Measurement function, the ranging accuracy can reach up to 3 mm. The modules include ultrasonic transmitters, receiver and control circuit.

- Working voltage: DC 5 V
- Working current: 15 mA
- Working frequency: 40 Hz
- Max range: 4 m
- Min range: 2 cm

FIGURE 4.4   Ultrasonic device.

- Measuring angle: 15 degree
- Trigger input signal: 10uS TTL pulse
- Echo output signal: input TTL lever signal and the range in proportion
- Dimension: 45*20*15 mm
- Test distance: (high level time × velocity of sound (340M/S) / 2

1 x Gyroscope Module MPU6050

- Digital-output X-, Y-, and Z-axis angular rate sensors (gyroscopes) with a user-programmable full-scale range of ±250, ±500, ±1000, and ±2000°/sec.
- External sync signal connected to the FSYNC pin supports image, video, and GPS synchronization.
- Integrated 16-bit ADCs enable simultaneous sampling of gyros.
- Enhanced bias and sensitivity temperature stability reduces the need for user calibration.
- Improved low-frequency noise performance.
- Digitally-programmable low-pass filter.
- Gyroscope operating current: 3.6 mA.
- Standby current: 5 μA.
- Factory calibrated sensitivity scale factor.
- User self-test.

1 x Accelerometer Module s MPU6050

- Digital-output triple-axis accelerometer with a programmable full scale range of ±2 g, ±4 g, ±8 g and ±16 g.
- Integrated 16-bit ADCs enable simultaneous sampling of accelerometers while requiring no external multiplexer.

FIGURE 4.5 Accelerometer device.

- Accelerometer normal operating current: 500 μA (Fig. 4.5).
- Low power accelerometer mode current: 10 μA at 1.25 Hz, 20 μA at 5 Hz, 60 μA at 20 Hz, 110 μA at 40 Hz.
- Orientation detection and signaling.
- Tap detection.
- User-programmable interrupts.
- High-G interrupt.
- User self-test.
- Two encoders.

1 x Servo Module SG90 (Fig. 4.6).

- Weight: 9 g
- Dimension: 22.2 × 11.8 × 31 mm approx.
- Stall torque: 1.8 kgf · cm
- Operating speed: 0.1 s/60 degree
- Operating voltage: 4.8 V (∼5 V)
- Dead band width: 10 μs
- Temperature range: 0−55 °C
- Approximately 180 degrees of rotation, 90 in each direction

1 x Arduino UNO Module (Fig. 4.7).

- Microcontroller: ATmega328
- Operating voltage: 5 V
- Input voltage (recommended): 7−12 V

FIGURE 4.6   Servo device.

FIGURE 4.7   Arduino Uno device.

FIGURE 4.8 Extension shield v5 device.

- Input voltage (limits): 6–20 V
- Digital I/O pins: 14 (of which 6 provide PWM output)
- Analog input pins: 6
- DC current per I/O Pin: 40 mA
- DC current for 3.3 V Pin: 50 mA
- Flash memory: 32 kb (ATmega328) of which 0.5 kb used by bootloader
- SRAM: 2 kb (ATmega328)
- EEPROM: 1 kb (ATmega328)
- Clock speed: 16 MHz

1 x Extension V5 Module (Fig. 4.8).

- I2C interface
- Bluetooth module communication interface
- SD card module communication interface
- APC220 wireless RF modules communication interface
- RB URF V1.1 ultrasonic sensors interface
- 12864 LCD serial and parallel interface
- 32 servo controller interfaces
- Dimensions: 57 × 57 × 18.5 mm
- Weight: 23 g

FIGURE 4.9   Bridge H L298N device.

1 x Bridge H L298N Module (Fig. 4.9).

- Operating voltage: 4–35 V
- Motor controller L298N, drives 2 DC motors or 1 stepper motor
- Max current 2 A per channel or 4 A max
- Free running stop and brake function
- Chip: ST L298N
- Logic power supply: 5 V
- Max power: 25 W
- Weight: 35 g
- Size: 55 mm × 60 mm × 30 mm
- Storage temperature: −25 to +135

2 x Batteries

- Generic 3800 mh 3.7 V Li-ion (Fig. 4.10)

Others

- Generic 4 motors
- Generic 4 wheels

4.2.6   MARCOS.COFRAM Handshake

The handshake between MARCOS.COFRAM and the arduino-based robot is required to negotiate the way both sides will perform during the current performance.

The fields' format is defined through a GUI interface that enables an administrator to easily manage the sequence and values expected from both sides.

FIGURE 4.10 Battery.

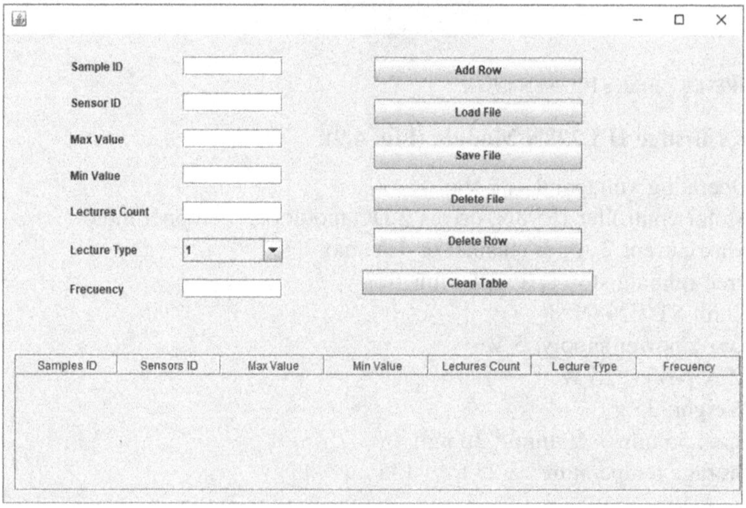

FIGURE 4.11 Handshake administrator interface.

Fig. 4.11 shows a screenshot of the Handshake administrator.

The file built by the application has comma separated values records (CSV) with the following fields:

Internal-ID
PIN-controller: with format AAAAA MMMM NNNN
    with AAAA being the model of the arduino device
    MMM factory or firm
    NNNN sequenced number
Number of data in the vector
Type of data

Init handshake message	DQ_MSJ_INIT

for each sensor	

Init id device	DQ_MSJ_ID_DEVICE
id device	(unit)

Init sensor sign	DQ_MSJ_ID_SENSOR
id sensor	(unit)
lectures amount	(unit)
min val	(double)
max val	(double)

check if readed	DQ MSJ ACK

FIGURE 4.12   Handshake interface.

Steps to establish the communication between MARCOS y COFRAN:

1. Send at the startup—Sensing or searching the devices in the robot.
   Figs. 4.12 and 4.13 show handshake interface.
2. With the device's metadata, the environmental information is gathered.

4.3 TESTING

Codelets that mimic bacteria behavior were added to the original code-rack prior to performing any tests. This will here be referred to as "Bacteria behavior" while the traditional consciousness system that evolves progressively from bottom knowledge will be referred to as "Consciousness behavior." Both cases use the same consciousness-based system (the COFRAM framework), but in the first case, a set of codelets mimic bacteria infections in order to produce new codelets with additional abilities. It is important to note that with plain consciousness behaving there are no means that this "extra" behavior can be automatically acquired. This paper aims to introduce, exhibit, and compare both configurations.

It is important to note that bacteria produce codelets with extra flexibility since its paradigm is quite different: adaptation through evolution and interaction with host. On the other hand, codelet creation and mutation (when it exists) in traditional consciousness is the result of mere learning by

```
 1  #include "Handshake.h"
 2  void Handshake::reset() { asm volatile (" jmp 0");}
 3  int Handshake::error_count = 0;
 4  DeviceQueue* Handshake::dq;
 5  DeviceQueue* Handshake::getDeviceQueue() {
 6      if (!Handshake::dq) {
 7          Handshake::dq = new DeviceQueue;
 8      }
 9      return Handshake::dq;
10  }
11  bool Handshake::init() {
12      int maxLoop;
13      DevicesLoader vel_info = DevicesLoader("");
14      maxLoop = HS_MAX_ERRORS;
15      while (Handshake::getDeviceQueue()->registerDevices(vel_info) != OK &&
16          --maxLoop >= 0) {};
17
18      if (maxLoop < 0) {
19          Handshake::reset();
20      }
21      return true;
22  }
23  bool Handshake::sensorsLoop() {
24      if (Handshake::getDeviceQueue()->sendData() != OK) {
25          Handshake::error_count++;
26      } else {
27          Handshake::error_count = 0;
28      }
29      if (Handshake::error_count > HS_MAX_ERRORS) {
30          Handshake::error_count = 0;
31          Handshake::reset();
32      }
33      return true;
34  }
```

```
 1  #ifndef HANSHAKE_H
 2  #define HANSHAKE_H
 3
 4  #include <Arduino.h>
 5  #include "DeviceQueue.h"
 6  #include "DevicesLoader.h"
 7
 8  #define HS_MAX_ERRORS 10
 9
10  class Handshake {
11      public:
12          static bool init();
13          static bool sensorsLoop();
14      private:
15          static void reset();
16          static int error_count;
17          static DeviceQueue* dq;
18          static DeviceQueue* getDeviceQueue();
19          Handshake();
20  };
21  #endif
```

FIGURE 4.13 Handshake interface.

reinforcement (see LIDA [30] for more details). That is a good approach when precision and consciousness do not have severe time restrictions; but this is not always the case, especially with autonomous mobile robots.

The remainder of this section presents three tests performed with bacteria and consciousness behavior. Every time, a summary of the main statistical results and explanatory figures are introduced, as well as a short interpretation.

4.3.1 Dataset

Three tests with different world-maps, produced a total of six datasets (two samples for each test, one for bacteria and the other for consciousness). Both datasets have the following variables:

- Timestamp: a sequence number to denote every step in the processing.
- clockTic: the ID of the tic when the clock advances. This is a virtual clock that serves to make evident which steps are to be performed during the same tic. Many of them will be implemented in parallel threads.
- Component: signature of the function in Java without parameters.
- Activity: task performed by the component.
- Detail: extra information for the activity.

All the testing was performed to evaluate the resolution ability of the codelet set, so the system keeps running until it reaches a strategy and declares it as the winner. There is no feedback or further reinforcement, since it would be more difficult to evaluate which part of the strategy was evolved by the codelets' quality and diversity, and which has evolved due to feedback (trial and error).

This approach is intended to overcome the codelet diversity in consciousness systems for autonomous mobile robots, avoiding the long evolution of the conscious until getting to the best ability to react to the environment. This is important since such robots need to react in real time to stimuli, while being flexible enough to avoid obstacles with smart strategies.

4.3.2 Test 1

The first test has one linear obstacle, as can be seen in Fig. 4.14. This constitutes the simplest test of the entire test set.

Fig. 4.15 shows the performance with and without bacteria-like behavior.

The light line shows the current obstacle position (as it was detected during robot displacement). The dark line is the real trace performed.

On the right, there is a compound image with both tests. As can be seen, they are almost the same except for a couple of dots. This is because the codelets are working in both cases using the same world-map and the resolution is quite simple.

22,752 records were collected for the test with bacteria behavior, and 11,463 records were collected for traditional consciousness behavior. The difference is due to the elevated variety and productivity of bacteria compared to the traditional consciousness system.

4.3.2.1 Global Behavior Performance

The first step is to compare the system's performance with and without bacteria. The descriptive stats are detailed in Table 4.1.

FIGURE 4.14 World-map for test 1.

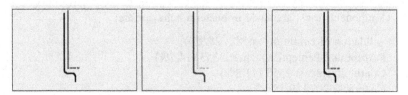

FIGURE 4.15 Test with (left) and without (middle) bacteria and with both (right).

TABLE 4.1 Descriptive Stats With and Without Bacteria

Variable	Bacteria	Consciousness
#activities performed	27	22
#clock tics	14	12
Most frequent activity	Created	Created
Less frequent	Winner, run	Winner, run
#components activated	36	26
Most frequent component	Percept	Percept

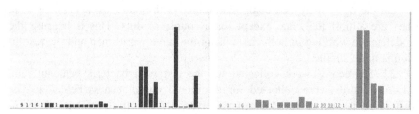

FIGURE 4.16 Components activation with bacteria (black) and traditional consciousness (gray).

Table 4.1 is a summary of the detailed descriptive statistics obtained with Infostat (c). As can be seen from the table, there are more "activity codelets" and "components activated" in Bacteria than in pure Consciousness. This is due to the bigger effort in building new strategies required by bacteria. But from the number of "clock ticks," it is clear that the system does not implement all the alternatives. The reason is simple: the system knows in advance which strategy is better and does not need to test them. The advantage is significant (14 against 251 for consciousness) which derives from the fact that bacteria criteria adds extra-powerful abstraction levels to the codelets in the coderack from the beginning, showing a good improvement in the performance.

In a more visual way, Figs. 4.16 and 4.17 show the histogram comparison for bacteria (black) and consciousness (gray) frequencies of component activation during the activity performed. Detailed frequencies are in Appendix A.

Components most activated in bacteria behavior are:

- validation of continuity: 6582 (28.9%)
- Percept and PerceptLauncher: 3353 (14.7%)
- ContinuityEdges: 2553 (11.2%)
- Coderack: 2079 (9.1%)
- Sensing: 1141 (5%)

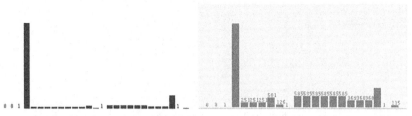

FIGURE 4.17   Activity performed bacteria (black) and traditional consciousness (gray).

The rest of the components remain under the threshold of 469 activations (2%).

Components activated in consciousness behavior are:

— Percept: 3510 (30.6%)
— Percept launcher: 3510 (30.6%)
— Sensing: 1105 (9.6%)
— Coderack: 1000 (8.7%)

The rest of the components remain under the threshold of 349 (3%).

From the activity percentages it can be said that the distribution of activities is fewer in Bacteria than in Consciousness, but in the first case dominates the evaluation of shapes (validation of continuity) while in the second there are more sensing and percept manipulation. That is due to the need to reevaluate the strategy many times because it is not sufficient to solve the trajectory determination.

An interesting matter is that Coderack (the set of codelets available) is being accessed more frequently in Bacteria. That is because there is a need to evaluate alternative codelets not present in the case of pure consciousness.

Here the efficiency of the bacteria system is made evident since in this case there is mainly one activity which is "to create," with 59.8% of the records in the log (13,609). That is because many different codelets are intensively being created and destroyed during exploration of strategy options (destruction of the codelets was not logged). In the case of consciousness, a similar scenario takes place since the creation activity dominates, but just with 37.8% of the records (4336 instances). This means that most of the activity is spread in many concomitant activities, but not by finding new strategies. Therefore exploration of approaches for the robots displacement becomes broader and, thus, more flexible.

4.3.2.2   Strategy for Resolution

This section compares the approach for finding a strategy with and without bacteria. In order to do this, the dataset was reduced, taking out all the records whose activity are other than DISTANCE, DISTANCE L, DISTANCE R, and CodeletExecution. This was done to evaluate only those

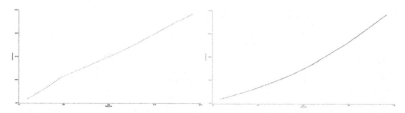

FIGURE 4.18 Dispersion diagram: Timestamp versus clockTick for bacteria (left) and consciousness (right).

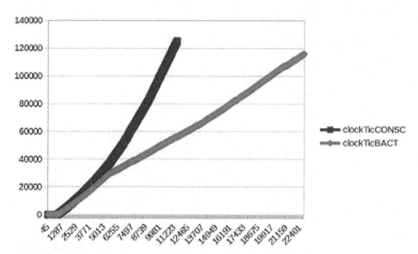

FIGURE 4.19 Comparison of clockTic evolution for bacteria and consciousness.

activities related to strategy resolutions and not world-map construction, codelet creation, and other additional tasks. This results in 3220 records for the bacteria test, and 2105 records for consciousness.

The dispersion diagram of timestamp (x) and clockTic (y), partitioned by activity (Codelet execution) is shown in Fig. 4.18 for bacteria (left) and for consciousness (right). Fig. 4.19 shows the superposition of both curves and it is clear that the number of strategies built are greater for bacteria than for consciousness.

Fig. 4.18 shows a different behavior in the activity for both approaches: more intensive for bacteria than for consciousness. In Fig. 4.6 clockTicCONSC denotes the evolution of tics when consciousness is running, while clockTicBACT stands for bacteria. For consciousness, the tics are quicker and the system ends earlier because the number of codelets in bacteria are higher and the decision takes more time. In the first case, the number of codelets are limited and represent global approaches (don't have time to learn and enhance strategy).

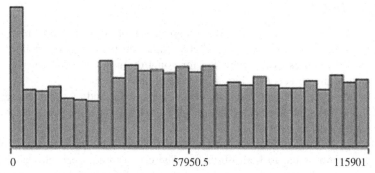

FIGURE 4.20   Comparison number of clockTics for bacteria.

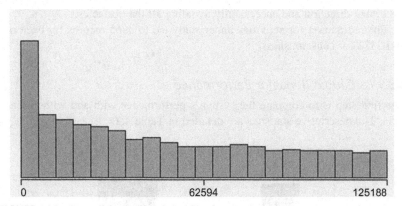

FIGURE 4.21   Comparison number of clockTics for consciousness.

Figs. 4.20 and 4.21 show a comparison between bacteria and consciousness frequencies, respectively, related to different strategies used in both cases.

In the case of bacteria, there are 232 different strategies created, while 251 strategies are created for consciousness. The first frequency in both cases for bacteria remains for a dummy strategy that is being used while the robot is performing its set-up and initial codelet preparation, before the bacteria starts its activity.

It is evident that the number of different strategies is higher for bacteria than for consciousness. So, it becomes clear that there is extra flexibility in the solution exploration along with more alternatives and possibilities to try. Therefore the system can explore and use different approaches each time. For consciousness, after an initial intensive search, the system calms down. For bacteria, there is an initial activity level that increases and afterwards it remains rather similar. Concretely, after an initial sensing and codelet startup, there are many other codelets, fired afterwards, that keep on looking for an improvement of the initial proposal of the strategy.

4.3.3 Test 2

To evaluate a more complex world-map, a set of obstacles was inserted. Although they do not constitute a continuous obstacle, the distribution is such that it requires extra evaluation for going through a global concave distribution of the objects (see Fig. 4.22 dotted line).

As for the previous test, the robot runs with and without bacteria. Fig. 4.23 shows the traces (left image for bacteria, right trace just consciousness as it naturally evolves).

In this scenario, the traces have more differences. For consciousness, the distance is not equal to both obstacles. There is a trend to go closer to the left obstacle. The strategy for bacteria seems to be better since it keeps in the middle of both. From a robotics perspective, it is a clear improvement since the distance to the obstacle is being kept safer without losing the original global direction and successfully avoiding all the obstacles.

Filtered dataset for activities under study led to 3965 records for bacteria, and 2372 for consciousness.

4.3.3.1 Global Behavior Performance

The first step is to compare the system's performance with and without bacteria. The descriptive statistics are detailed in Table 4.2.

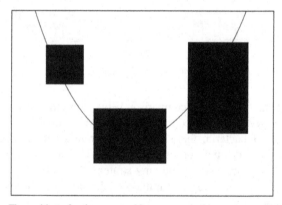

FIGURE 4.22 Three objects for the new world-map, arranged in a concave distribution.

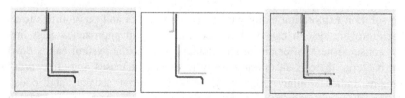

FIGURE 4.23 Trace with bacteria (left) and just consciousness (middle) and both (right).

TABLE 4.2 Descriptive Statistics With and Without Bacteria

Variable	Bacteria	Consciousness
#activities performed	27	22
#clock tics	15	13
Most frequent activity	Created	Created
Less frequent	Winner, run	Winner, run
#components activated	36	25
Most frequent component	ci2s.marcos.concepts.links.cell. Adjacency	Percept

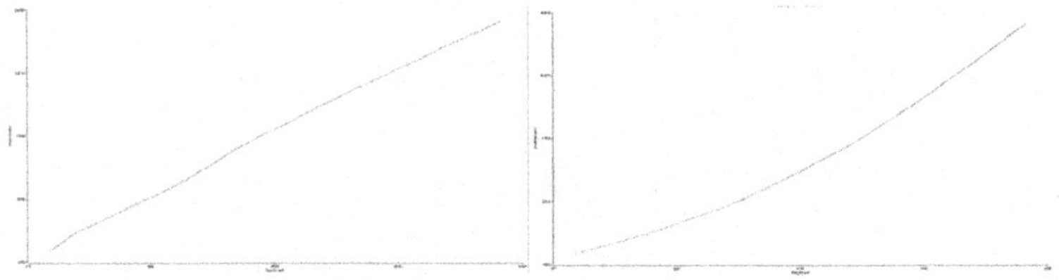

FIGURE 4.24 Comparison of clockTic evolution for bacteria (left) and consciousness (right).

The key component for bacteria in this test is the adjacency concept that reflects the search for the coherence between sensed cells. For consciousness, the elaboration was replaced by the permanent requirements to get the same information from the environment using the sensing mechanism (percept component).

To sum up, bacteria develops a strategy without requiring further perception from the environment. It elaborates abstract concepts (adjacency, linearity, etc.) for the set of information gathered from percepts.

4.3.3.2 Strategy for Resolution

This section compares the approach for finding the strategy. To do so, the dataset was reduced, taking out all the records of those whose activity are other than DISTANCE, DISTANCE L, DISTANCE R, and CodeletExecution. This was done in order to evaluate only the activities related to strategy resolutions and no world-map construction, codelet creation, and other additional tasks. That leads to a total of 3965 records for the bacteria test, and 2372 records for consciousness.

Fig. 4.24 shows dispersion plots for both approaches and Fig. 4.25 compares the tics taken for both approaches during the test.

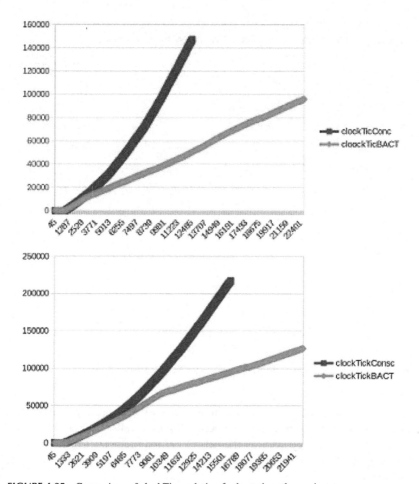

FIGURE 4.25 Comparison of clockTic evolution for bacteria and consciousness.

This test keeps the same trends from Test 1, even though the world-map is more complex. It is interesting to compare the codelet's activation frequency histogram of Fig. 4.26: the activity is now different for consciousness but remains the same for bacteria.

This clearly indicates that bacteria behave similarly even when the task is more complex, denoting the efficiency and simplicity of this method.

4.3.4 Test 3

This final test aims to evaluate the concave- and convex-shaped obstacles. Fig. 4.27 shows the new world-map.

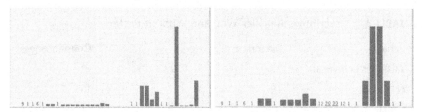

FIGURE 4.26 Comparison of components activated for bacteria (left) and consciousness (right).

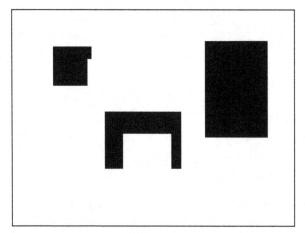

FIGURE 4.27 World-map with concave and convex shaped obstacles.

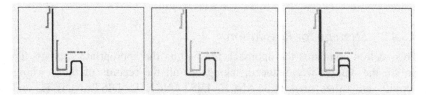

FIGURE 4.28 Test for bacteria (left), consciousness (middle) and the comparison (right).

The test results in the traces indicated in Fig. 4.28 for bacteria (left), consciousness (middle) and the comparison (right).

As can be noted, both strategies differ more than in the previous tests. The security margin increases the trace's efficiency.

4.3.4.1 Global Behavior Performance

The first step is to compare the system's performance with and without bacteria. The descriptive statistics are detailed in Table 4.3.

TABLE 4.3 Descriptive Statistics With and Without Bacteria

Variable	Bacteria	Consciousness
#activities performed	27	22
#clock tics	15	13
Most frequent activity	Created	Created
Less frequent	Winner, run	Winner, run
#components activated	37	25
Most frequent component	ci2s.marcos.concepts.links.cell. Adjacency	Percept

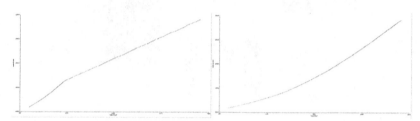

FIGURE 4.29 Comparison of clockTic evolution for bacteria (left) and consciousness (right).

The table is very similar to Table 4.2, for Test 2, confirming the trend observed for that test.

4.3.4.2   Strategy for Resolution

This section compares the approach for finding the appropriate strategy. To do so, the dataset was reduced, taking out all the records of those whose activity are other than DISTANCE, DISTANCE L, DISTANCE R, and CodeletExecution. This is done in order to evaluate only the activities related to strategy resolutions and no world-map construction, codelet creation, and other additional tasks. That leads to a total of 5550 records for the bacteria test, and 3272 records for codelets.

Fig. 4.29 shows dispersion plots for both approaches and Fig. 4.30 compares the tics taken for both approaches during the test.

This indicates that bacteria behaves similarly even when the task is more complex.

It keeps the same trends as for Tests 1 and 2, even though the world-map is more complex. Comparing Fig. 4.30 and Figs. 4.24, 4.25, the activity remains the same for both approaches (Fig. 4.31).

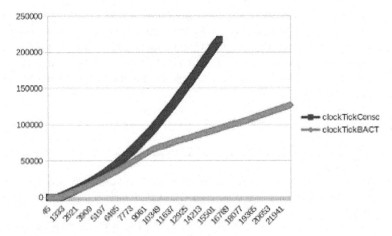

FIGURE 4.30   Comparison of clockTic evolution for bacteria.

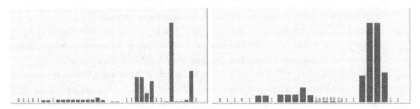

FIGURE 4.31   Comparison of components activated for bacteria (left) and consciousness (right).

4.4   CONCLUSIONS AND FUTURE WORK

This chapter presented the basis of a new approach that intends to add flexibility, efficiency, and creativity to the decisions taken by consciousness when applied to autonomous mobile robots. Three tests, in increasingly obstacle-avoidance complexity, were performed in order to compare traditional codelets versus the ones created with bacteria. The focus is on "activity codelets" and "components activated," given that the bacteria paradigm provides adaptation through evolution and interaction.

These tests evaluate how bacteria produce more "activity codelets" and is more "component activated" than the plain consciousness system, adding an extra-powerful abstraction level to the codelets showing true improvement in the performance. It is also evident from the results that the diversity of strategies is higher for bacteria (making it clear that there is an extra flexibility in the exploration of the solution). At the same time, this gives the possibility of processing complex obstacle distributions more efficiently. In conclusion, bacteria-based consciousness robots behave very well in tasks with different degrees of complexity, broadening the scope a robot can be put into.

As for future work, there are many important things to work on. Among others to implement temporality to the codelets colony in order to achieve the auto-regulation according to the current strategy, research and implement new infections, tune the communication between codelets to imitate the behavior in a bacterial colony, and test new strategies to overcome abrupt level differences in the path.

REFERENCES

[1] R. Reddy, No. 3. May/June Robotics and Intelligent Systems in Support of Society, vol. 21, Carnegie Mellon University, 2006.

[2] D.L. De Luise, L. Rancez, N. Biedma, B. De Elía, D.L. Colodrero, J. Mendez, Odometer errors for a robot controlled by consciousness, 2016 IEEE Biennial Congress of Argentina (ARGENCON), June 15–17, IEEE, Buenos Aires, Argentina, 2016.

[3] F. Penizzotto, E. Slawiñski, V. Mut, Laser radar based autonomous mobile robot guidance system for olive groves navigation, IEEE Lat. Am. Trans. 13 (5) (2015) 1303–1312.

[4] A. Di Fava, M. Satler, P. Tripicchio, Visual navigation of mobile robots for autonomous patrolling of indoor and outdoor areas, 2015 23rd Mediterranean Conference on Control and Automation (MED), June 16–19, IEEE, Torremolinos, Spain, 2015.

[5] A. Chand, S. Yuta, Navigation strategy and path planning for autonomous road crossing by outdoor mobile robots, The 15th International Conference on Advanced Robotics Tallinn University of Technology Tallinn, Estonia, June 20–23, IEEE, Tallinn, Estonia, 2011.

[6] Y. Chiu, N. Shiroma, H. Igarashi, N. Sato, M. Inami, F. Matsuno, FUMA: environment information gathering wheeled rescue robot with one-DOF arm, Proceedings of the 2005 IEEE International Workshop on Safety, Security and Rescue Robotics, June 6–9, IEEE, Kobe, Japan, 2005.

[7] K. Shimazaki, T. Nagao, Semantic segmentation considering location and co-occurrence in scene, 2015 IEEE 8th International Workshop on Computational Intelligence and Applications November 6–7, IEEE, Hiroshima, Japan, 2015.

[8] Y. Zhuang, W. Wang, H. Chen, K. Zheng, 3D scene reconstruction and motion planning for an autonomous mobile robot in complex outdoor scenes, Proceedings of the 2010 International Conference on Modelling, Identification and Control, July 17–19, IEEE, Okayama, Japan, 2010.

[9] S. Iwata, S. Sahashi, T. Hasegawa, Verification of visual attitude control for outdoor autonomous mobile robots, 2013 10th International Conference on Ubiquitous Robots and Ambient Intelligence (URAI), IEEE, Jeju, South Korea, 2013.

[10] H. Kawamura, S. Iwata, S. Sahashi, T. Hasegawa, Visual attitude control using a virtual barycenter of a quadrangle that constructed from feature points for outdoor autonomous mobile robots, 2012 International Symposium on Micro-NanoMechatronics and Human Science (MHS), IEEE, Nagoya, Japan, 2012.

[11] J.E. Jung, K.S. Lee, H.G. Park, Y.H. Koh, J.I. Bae, M.H. Lee, Vision-based obstacle avoidance system for autonomous mobile robot in outdoor environment, International Conference on Control, Automation and Systems 2010, Oct. 27–30, IEEE, KINTEX, Gyeonggi-do, South Korea, 2010.

[12] U.Z. Sanal, A.M. Erkmen, I. Erkmen, Logical sensing and intelligent perception based on rough sets and dynamic decay adjustment learning, Proceedings of the 2002 IEEE International Conference on Robotics 8 Automation, May 11–15, IEEE, Washington, DC, 2002.

[13]   U. Witkowski, M. El-Habbal, S. Herbrechtsmeier, A. Tanoto, J. Penders, L. Alboul, et al., Ad-hoc Network Communication Infrastructure for Multirobot Systems in Disaster Scenarios, Proceedings of IARP/EURON Workshop on Robotics for Risky Interventions and Environmental Surveillance (RISE 2008), Benicassim, Spain, 2008.

[14]   A. Denker, M.C. Iseri, Design and implementation of a semi-autonomos mobie search and rescue robot: SALVOR, 2017 International Artificial Intelligence and Data Processing Symposium (IDAP), September 16−17, IEEE, Malatya, Turkey, 2017.

[15]   A. Konde-Deshmukh, J. Doshi, J. Kothari, M. Kawa, V. Wadhe. Disaster rescue robot. Int. J. Electr. Electron. Res. 4(1), 120−124, ISSN 2348-6988 (online), Available at: <www.researchpublish.com>

[16]   A.R. Krishna, G.S. Bala, A.S.N. Chakravarthy, B.B.P. Sarma, G.S. Alla, Design of a rescue robot assist at fire disaster, Int. J. Comput. Appl. (0975−888) 47 (10) (2012).

[17]   S. Bag, P. Bhowmick, G. Harit, Detection of structural concavities in character images—a writer-independent approach, Indo-Japanese Conference on Perception and Machine Intelligence, Lecture Notes in Computer Science, 7143, 2012.

[18]   A. Hussein, P. Marín-Plaza, D. Martín, A. de la Escalera, J.M. Armingol, Autonomous off-road navigation using stereo-vision and laser-rangefinder fusion for outdoor obstacles detection, 2016 IEEE Intelligent Vehicles Symposium (IV), June 19−22, IEEE, Gothenburg, Sweden, 2016.

[19]   S. Muhammad, D. Nardi, K. Ohno, S. Tadokoro, Environmental sensing using millimeter wave sensor for extreme conditions, 2015 IEEE International Symposium on Safety, Security, and Rescue Robotics (SSRR), October 18−20, IEEE, West Lafayette, IN, USA, 2015.

[20]   H. Chae, Christiand, S. Choi, W. Yu, J. Cho. Autonomous navigation of mobile robot based on DGPS/INS sensor fusion by EKF in semi-outdoor structured environment. 2010 IEEE/RSJ International Conference on Intelligent Robots and Systems, October 18−22, 2010, IEEE, Taipei, Taiwan.

[21]   M.N. Kiyani and M.U.M. Khan. A prototype of search and rescue robot. 2016 2nd International Conference on Robotics and Artificial Intelligence (ICRAI), November 1−2, IEEE, Rawalpindi, Pakistan.

[22]   L. Kuhnert, B. Meier, K.-D. Kuhnert, Probabilistic approach to self-localization for autonomous mobile outdoor robotics based on hybrid map knowledge, Proceedings of the 16th International IEEE Annual Conference on Intelligent Transportation Systems (ITSC 2013), October 6−9, IEEE, The Hague, The Netherlands, 2013.

[23]   M. Zhou, W. Gao, Multi-sensor data acquisition for an autonomous mobile outdoor robot, 2011 Fourth International Symposium on Computational Intelligence and Design, October 28−30, IEEE, Hangzhou, China, 2011.

[24]   K.-D. Kuhnert, Software architecture of the autonomous mobile outdoor robot AMOR, 2008 IEEE Intelligent Vehicles Symposium, June 4−6, IEEE, Eindhoven, Netherlands, 2008.

[25]   K.-D. Kuhnert, Concept and implementation of a software system on the autonomous mobile outdoor robot AMOR, 2008 IEEE International Conference on Industrial Technology April 21−24, IEEE, Chengdu, China, 2008.

[26]   D. López De Luise, L. Rancez, N. Biedma and B. De Elía. A conscious model for autonomous robotics: statistical evaluation, 2015, Int. J. Adv. Intell. Paradig. https://doi.org/10.1504/IJAIP.2015.070331

[27]   R. Kondadadi, S. Franklin, Deliberative Decision Making in "Conscious" Software Agents.

[28] E. Ben-Jacob, Learning from bacteria about natural information processing, natural genetic engineering and natural genome editing, Ann. N. Y. Acad. Sci. 1178 (1) (2009) 78–90.

[29] M. Ahmed, M.R. Khan, M.M. Billah, S. Farhana. Walking hexapod robot in disaster recovery: developing algorithm for terrain negotiation and navigation., New Advanced Technologies, A. Lazinica (Ed.), InTech, ISBN: 978-953-307-067-4, Available from: http://www.intechopen.com/books/new-advanced-technologies/walking-hexapod-robot-in-disaster-recoverydeveloping-algorithm-for-terrain-negotiation-and-navigation.

[30] S. Franklin, U. Ramamurthy, S. D'Mello, L. McCauley, A. Negatu, R. Silva, et al., LIDA: a computational model of global workspace theory and developmental learning, AAAI Fall Symposium on AI and Consciousness: Theoretical Foundations and Current Approaches, AAAI, Arlington, VA, 2007.

FURTHER READING

B.J. Baars, A Cognitive Theory of Consciousness, Cambridge University Press, 1998.

B.J. Baars, In the Theater of Consciousness: The Workspace of the Mind, Oxford University Press, 1997.

S. Bag, G. Harit, A medial axis based thinning strategy for character images, Proceedings of the second National Conference on Computer Vision, Pattern Recognition, Image Processing and Graphics (NCVPRIPG), arXiv.org, Jaipur, India, 2011.

R. Bostelman, T. Hong, R. Madhavan, T. Chang, H. Scott, Performance analysis of an autonomous mobile robot mapping system for outdoor environments, 2006 9th International Conference on Control, Automation, Robotics and Vision, December 5–8, IEEE, Singapore, 2006.

D. Dennet, M. Kinsbourne, Time and the observer: the where and when of consciousness in the brain, Behav. Brain Sci. 15 (2) (1992) 183–201. <http://www.sc.ehu.es/sbweb/fisica_/cinematica/rectilineo/rectilineo/rectilineo.html>.

H. Matute, C.P. Cubillas, P. Garaizar, Learning to infer the time of our actions and decisions from their consequences, Conscious. Cogn. 56 (2017) 37–49.

New Robot Strategy. Ministry of Economy, Trade and Industry of Japan.

Q. Qiu, J. Han, Function sector based real-time autonomous navigation for outdoor mobile robots equipped with laser scanners, 2009 IEEE International Conference on Control and Automation, December 9–11, IEEE, Christchurch, New Zealand, 2009.

I. Ulrich, I. Nourbakhsh, Appearance-based obstacle detection with monocular color vision, Proceedings of the AAAI National Conference on Artificial Intelligence, July 30–August 03, AAAI Press, Austin, TX, 2000, pp. 866–871.

R.S. Xavier, N. Omar, L.N. de Castro, Bacterial colony: information processing and computational behavior, 2011 Third World Congress on Nature and Biologically Inspired Computing, October 19–21, IEEE, Salamanca, Spain, 2011.

Z. Xuhui, D. Runlin, L. Yongwei, VR-based remote control system for rescue detection robot in coal mine, 2017 14th International Conference on Ubiquitous Robots and Ambient Intelligence (URAI), June 28–July 1, IEEE, Jeju, South Korea, 2017.

Y. Zhuang, S. Yang, X. Li, W. Wang, 3D-Laser-based visual odometry for autonomous mobile robot in outdoor environments, 2011 3rd International Conference on Awareness Science and Technology (iCAST), September 27–30, IEEE, Dalian, China, 2011.

Chapter 5

Analysis of Transform-Based Compression Techniques for MRI and CT Images

Eben Sophia Paul[1] and J. Anitha[2]

[1]Department of ECE, Karpagam University, Coimbatore, India, [2]Department of Electronics and Communication, Karunya University, Coimbatore, India

Chapter Outline

5.1 INTRODUCTION

Telemedicine is the process of delivering or exchanging health-care information over distances. However, the size of medical images makes telemedicine process complex and impractical. There are various imaging modalities such as magnetic resonance imaging (MRI) and computed tomography (CT), and their complexity keeps increasing with increasing technology. The images generated using MRI or CT scans are usually reformatted and transformed into multiple planes. Hence, for a single patient a large number of images are generated for a single study. There are different means by which these images are viewed, including on a computer monitor, or transferred to a CD or DVD, or printed on film. In case of telemedicine application, they need to be stored in any of these format and has to be transmitted. Quality

Intelligent Data Analysis for Biomedical Applications. DOI: https://doi.org/10.1016/B978-0-12-815553-0.00005-7

compression is required at each stage of the process for easy storage and transmission and for the radiologist to analyze the images correctly and prepare diagnosis reports [1]. Image compression is the process of reducing the size of the image with or without degrading the image data. In the case of medical images, any loss of clinical information may lead to legal issues. Compressing a medical image is a complex task and requires lot of mental effort. The European Society of Radiology [2] has done research on irreversible (lossy) compression in radiology and has provided information about its impact on the use of irreversible image compression and on diagnostically acceptable irreversible compression. However, lossy compression is rarely allowed for preserving quality. Since lossless compression can contribute only up to 3:1 reduction in size [3], this further increases the complexity. But the use of irreversible compression also helps in reducing the storage and bandwidth requirement and improves radiology services [4]. They can also be used for compressing the nonregions of interest of the medical images. Region-based compression is required to induce high compression on nonregions of interest, thereby achieving improved performance [5−9].

Standard compression algorithms such as JPEG and JPEG2000 are widely used in the medical field. Both the methods are transform-based compression approaches. They convert image from one domain to another domain using image transforms and enables efficient coding of the transformed coefficients. They are computationally complex, but the compression achieved is higher compared to other methods. Adaptive transforms such as brushlet [10], bandelet [11], and directionlet [12] are not suitable for medical images because of their complexity. The basis function of these transforms changes with respect to the content of the image and makes it more complicated. Hence nonadaptive transforms such as wavelet [13], contourlet [14], and ripplet [15] are normally used for medical images. Even though there are many transform-based lossy compression algorithm available in the literature [16−20], there is a need for improvement. We also need to know which transform performs better for a specific modality of medical images.

A transformation alone does not produce compression and hence a preprocessing technique along with encoding techniques is required for enhanced compression. As a part of quantization several lossy coding algorithms can also be experimented with. In this chapter, we have used a symmetry-based preprocessing technique as in [21] for medical images reformatted in multiple planes. Here, instead of comparing the similarity within the image, we utilize the similarity between the consecutive images which are transformed in multiple planes. This approach is a complete lossless approach and the images which are preprocessed can be reconstructed without any loss of data. This is followed by transformation of the preprocessed image data. Then, Random Singular Value Decomposition (RSVD) [22] is used to efficiently decompose larger subbands in case of contourlet and ripplet transforms. Many lossy compression algorithms use SVD for obtaining high PSNR [23,24]. In the same way, RSVD is used for decomposing transformed subbands and reconstructing with good PSNR. Optimized mask-based prediction is used as part of the encoding procedures to capture the randomness and smoothness of the image data [25−28]. Multiple dataset images are used for comparison between different transforms and two different modalities of medical images such as MRI and CT images of different sizes. The results of 240 test images are presented and analyzed with relevance to the concept of different transformations that suits for different medical imaging modality for telemedicine applications.

5.2 PROPOSED METHODS

5.2.1 Overall Transform-Based Compression Model

The key operations of this encoder part are: (1) the preprocessing unit which captures the symmetry between consecutive sets of images through which the less complex image difference is encoded instead of the original image; (2) the transformation unit which transforms the image coefficients in spatial domain to achieve compression; and (3) the coding unit which codes the transformed coefficients by means of optimized prediction, random singular value decomposition, and uniform scalar quantization. The encoding algorithm used is the conventional arithmetic encoding technique which is the fastest and simplest. The overall block diagram is shown in Fig. 5.1. Work is carried out in all the blocks shown which are discussed next. Detail explanations of these techniques can be found in Refs. [29,30,31].

5.2.2 Preprocessing Unit

The images generated using MRI or CT scans are usually reformatted and transformed into multiple planes. Hence, for a single patient, the number of images generated for a single study is large in number. But, these images have more similarities between them which can be captured for efficient compression results. A step-by-step transformation of planes does not produce much change and, hence, their symmetry can be captured so that the difference can be encoded instead of the original image. The explanation can be given by means of the mathematical model discussed next.

Consider there are 20 clinical images and are represented as C1, C2,..., C20. The first image, C1, can be encoded as such, whereas the next image, C2, can be coded as (C2−C1) and so on. We can represent these series of long sums by means of sigma (Σ) which is the summation notation. The differences between the consecutive images are given by:

$$T_n = \sum_{n=2}^{20} C_n - D_n \qquad (5.1)$$

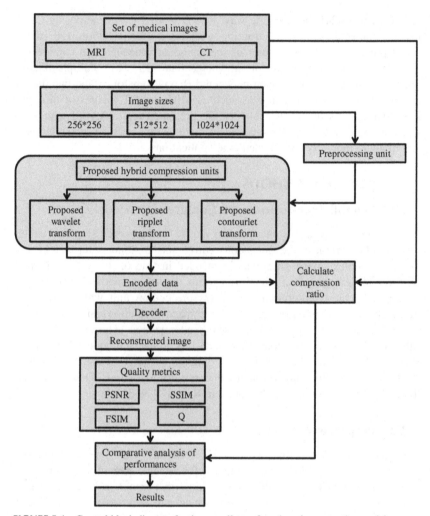

FIGURE 5.1 General block diagram for the overall transform-based compression model.

where $T_1 = C_1$ and D_n can be calculated as

$$D_n = \sum_{n=3}^{20} T_{n-1} + C_{n-2} \tag{5.2}$$

where $D_1 = 0$ and $D_2 = T_1$.

T_n is the difference image that will be encoded instead of the original image. This difference image will have very less entropy compared to the original image and can be encoded easily. After reconstruction, C_n, the original image can be retrieved using T_n and D_n as:

$$C_n = \sum_{n=1}^{20} Trec_n + D_n \tag{5.3}$$

where *Trec* is the reconstructed image and D_n can be calculated using Eq. (5.2). This is the proposed method by which a series of images which are of same modality and are continuous and direction and time variant can be compressed efficiently without any loss of information. Instead of compressing the consecutive images directly, the difference images are compressed and after decoding they are reconstructed and retrieved.

5.2.3   Optimized Prediction Unit

Prediction captures the randomness and smoothness of the image data by exploiting the redundancies. The proposed optimized prediction approach finds the exact threshold value for masking before prediction. Higher values are predicted using prediction functions and lower values are set to zero after prediction to avoid lower values resulting in higher prediction errors. This is a lossless procedure where the data can be regenerated by means of inverse prediction using the same prediction function. The optimization algorithm used here is the fastest and the least complex particle swarm optimization (PSO) technique. The proposed optimized prediction approach is summarized next.

1. Initialization

    Set the initial threshold value (population) to the values randomly selected around the mean value of the image data and initialize all the PSO parameters.
2. Masking

    Mask the value above the threshold value by means of binary masking. When the data is greater than the threshold value set the masked coefficient as 1 and use 0 for the others.
3. Predictive coding

    There are various prediction functions and the function which gives less entropy is selected. If the previous pixel value $a(n-1)$ of $a(n)$ is X, Y, and Z, then, $a_p(n)$ can be predicted as either X, Y, Z, $X + (Y - Z)$, $X + (Z - Y)/2$, $Y + (X - Z)/2$, $(Y + Z)/2$, $(X + Z)/2$, $(X + Y)/2$, or $(X + Y + Z)/3$. The prediction error is found by finding the difference between $a(n)$ and $a_p(n)$. This error value is then processed to the next level of compression.
4. Threshold updating

The threshold value can be optimized using PSO. The best value can be selected from the alternatives available through a rigorous selection process based on the PSO algorithm. They are used for finding near-optimal solutions and are very useful in image compression to find the best match for better results. The threshold value is selected to create a mask which makes the process of prediction easier. PSO has a wide range of applications, such as pattern recognition, determining intensity points, noise filtering, enhancement, classification, and segmentation. PSO reduces the search space for finding the self-similarities in the given image. In PSO, the particles are moved toward the "pbest" (Personal best) and the "gbest" (Global best) solution. The mask-based prediction approach effectively utilizes PSO for finding the exact threshold value (which provides high CR) for masking. Then the unmasked coefficients are predicted and the prediction errors are arithmetically encoded. The block diagram of the proposed PSO-based prediction model is shown in Fig. 5.2.

For PSO, the swarm size was initially set to 5 and positions of each particle are selected randomly around the mean value of the input coefficients. The fitness function here is the CR and the particle of the whole swarm, which gives higher CR, is considered as pbest and the gbest gets updated as the best of pbest after each iteration. Here, the user will specify the number of iterations in order to avoid time complexity introduced while setting the stopping criterion. After each iteration the position of the particles and velocity are updated for calculating the updated population for the next iteration.

The velocity (V) and population (P) can be updated using these equations:

$$V_{ij} = \left(I^*V_{ij}\right) + \left(c_{f1}{}^*r_{v1}\left(pbest - P_{ij}\right)\right) + \left(c_{f2}{}^*r_{v2}\left(gbest - P_{ij}\right)\right) \qquad (5.4)$$

$$P_{ij} = P_{ij} + V_{ij} \qquad (5.5)$$

where c_{f1} and c_{f2} are the correction factors set to 2, r_{v1} and r_{v2} are random values between 0 and 1; I is inertia set to 0.8; V is the particles' velocity;

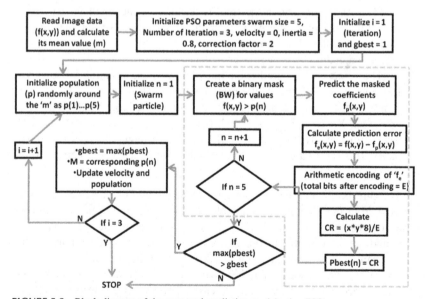

FIGURE 5.2   Block diagram of the proposed prediction model using PSO.

P is the current particles; i is the iteration number; and j is the particle position of the current iteration.

Repeat the process until the maximum compression ratio is found, or until the specified number of iterations is reached.

5.2.4   Wavelet Transform-Based Compression Unit

Each of the transform operations produce different sizes of subband coefficients. In the case of wavelet transforms, the subband sizes are equal to the size of the image while the number of subbands depends on the level of decomposition. The proposed wavelet transforms (PWT) based compression model is shown in Fig. 5.3.

Wavelet transform decomposes the image into wavelet coefficients of different frequencies. The filter used here is the db5 which is the simplest of wavelet filters. The decomposed images are predicted and quantized, and entropy is encoded. In the case of contourlet transform, the subband sizes are larger than the image size and keeps increasing with the increasing level of decomposition. So, in addition to prediction and entropy encoding, we have used singular value decomposition to further decompose the high frequency subband coefficients with less loss of information

5.2.5   Contourlet Transform-Based Compression Unit

The proposed contourlet transform (PCT) based compression approach is shown in Fig. 5.4.

The ripplet transform subbands are similar to contourlet subbands. The sizes are large since they provide directional information at different scales. So RSVD is used here to efficiently reduce the subband sizes and induce compression with less loss of information.

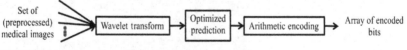

FIGURE 5.3   Proposed wavelet transform-based compression.

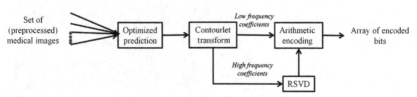

FIGURE 5.4   Proposed contourlet transform-based compression.

5.2.6 Ripplet Transform-Based Compression Unit

The proposed ripplet transform (PRT) based compression approach is shown in Fig. 5.5.

The Laplacian filter used is 9/7 wavelet filters and the directional filters used for high-pass bands are pkva ladder filters. The scaling variable c and d are set to 1 and 3, respectively, and the decomposition levels are calculated using c and d and are found to be [0, 1, 2, 2].

5.2.7 Arithmetic Coding Unit (Conventional)

Finally the encoding algorithm used is arithmetic encoding and the procedure is explained in Fig. 5.6 with a small example.

In Fig. 5.6, a small sequence (ACD) is encoded using the arithmetic coding technique to understand its operation. Initially the probability of occurrence of each symbol is calculated. Based on these probabilities, the probability range is calculated for each symbol as shown. In order to encode A, all the symbols are mapped within the probability range of A. Then to encode C, again, all the symbols are mapped within the new probability

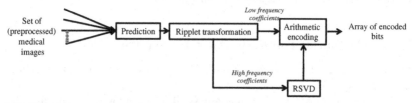

FIGURE 5.5 Proposed ripplet transform-based compression.

Symbols	Probability	Probability range
A	0.4	0-0.4
B	0.3	0.4-0.7
C	0.1	0.7-0.8
D	0.1	0.8-0.9
E	0.06	0.9-0.96
F	0.04	0.96-1

Sequence to be encoded: "ACD"

Encoded value can be any value between 0.312 and 0.316

FIGURE 5.6 Arithmetic encoding with example.

range of C and this applies for D as well. The final range of probability obtained (0.312−0.316) is sufficient to reconstruct the sequence. Any value between the ranges can be used for decoding the input sequence in the same way. This is a lossless encoding procedure which encodes the data based on the probability of occurrence of each symbol.

5.3   RESULTS AND DISCUSSION

In this section, we evaluate the performance of three different transform-based compression techniques to find the one best-suited for a specific modality of medical images. The performance also differs with respect to the variation in image sizes, which are also analyzed. The quality measures used for comparing the performance are Peak-Signal-To-Noise−Ratio (PSNR), Structure SIMilarity Index (SSIM), Feature SIMilarity Index (FSIM), and Universal Quality Index (Q). The distortion is measured by the PSNR in dB:

$$PSNR = 10 \log_{10}\left((255)^2/MSE\right) \tag{5.6}$$

where mean-square-error (MSE) is

$$MSE = \frac{1}{mn}\sum_{i=1}^{m}\sum_{j=1}^{n}\left|x_{ij} - y_{ij}\right| \tag{5.7}$$

MSE gives the error value between the reconstructed image (y) and the original image (x). The variables m and n are the sizes of the original image. The structure similarity is calculated using SSIM and is given by:

$$SSIM = \frac{\left(2\mu_f\mu_{\bar{f}} + C_1\right)(2\delta_{f\bar{f}} + C_2)}{\left(\mu_f^2 + \mu_{\bar{f}}^2 + C_1\right)(\delta_f^2 + \delta_{\bar{f}}^2 + C_2)} \tag{5.8}$$

where f is the original image and $\bar{f}$ is the reconstructed image, μ represents the average gray value, δ represents the variance, and C_1 and C_2 are constants to prevent unstable results.

The feature similarity between the original and the reconstructed image is captured using FSIM:

$$FSIM = \frac{\sum_{x\in\Omega}S_L(X).PC_m(X)}{\sum_{x\in\Omega}PC_m(X)} \tag{5.9}$$

$$where\ PC_m(x) = \max(PC_1(x), PC_2(x)) \tag{5.10}$$

$PC_1(x)$ and $PC_2(x)$ are the phase congruency of the reference image and the decompressed image, $S_L(x)$ is the similarity between both images at each location x, and Ω is the whole image's spatial domain.

The universal quality of the reconstructed image is measured using Q, which is the universal quality index and is given by:

$$Q = \frac{1}{M} \sum_{j=1}^{M} Q_j \qquad (5.11)$$

where M is the number of local regions covered by the sliding window and Q_j is the local quality index at the local region j which is given by:

$$Q_j = \frac{4\sigma_{xy}\bar{x}\bar{y}}{\left(\sigma_x^2 + \sigma_y^2\right)[\bar{x}^2 + \bar{y}^2]} \qquad (5.12)$$

where x and y are the original and the decompressed images, σ_x and σ_y are the estimates of the contrast between x and y.

These are some of the gray image quality measures based on the luminance component which are used for calculating the quality of the reconstructed image with comparison to the original image.

The Bits Per Pixel (BPP) which gives the bit rate is calculated as:

$$\text{BPP} = \frac{\text{Encoded bytes}^*8}{m^*n} \qquad (5.13)$$

The sample test images used for analysis are shown in Fig. 5.7. Five continuous set of images are shown here for each category of medical images. Similarly, 240 test images were used on the whole set to analyze the results. The results of the preprocessing are presented next followed by the compression performance of the different transform-based techniques and their comparison.

5.3.1 Preprocessing Unit Results

Each image is preprocessed to capture the symmetry between the previous image and the image that has to be processed. The set of continuous reformatted images are shown in Fig. 5.7. These images are MRI and CT scanned images which are transformed in multiple planes. The similarity between two consecutive images can be captured efficiently as a part of preprocessing. This helps in reducing the source entropy of the original image, as shown in Table 5.1.

Here, H1 is the entropy of the original image and H2 is the entropy of the preprocessed image. For real-time images the entropy after preprocessing decreases on a larger scale compared to online database images due to the higher level of symmetry between the real-time images. This preprocessing helps in reducing the coding complexity and the bit rate, thereby enabling easy and fast compression of a continuous set of medical images.

5.3.2 Overall Compression Results

2D MRI and CT images of sizes 256×256, 512×512, and 1024×1024 are used for comparing the results of wavelet, contourlet, and ripplet

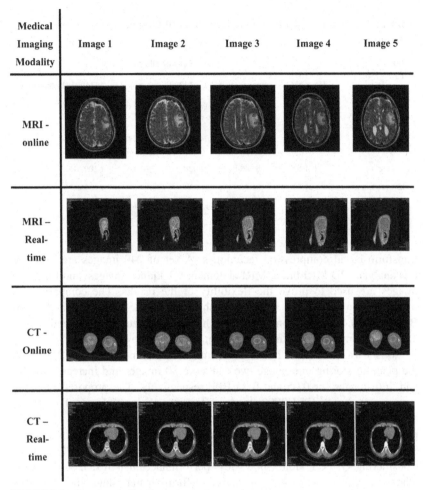

FIGURE 5.7   Sample set of test images.

TABLE 5.1 Entropy Comparison of the Original and the Preprocessed Images

Images	Image2		Image3		Image4		Image5	
	H1	H2	H1	H2	H1	H2	H1	H2
MRI-online	5.76	**3.24**	5.87	**3.49**	5.95	**2.61**	6.11	**3.54**
MRI-real-time	3.71	**0.49**	3.72	**0.76**	3.76	**1.46**	3.88	**1.01**
CT-online	6.5	**5.31**	6.55	**5.36**	6.49	**5.41**	6.56	**5.41**
CT-real-time	2.09	**0.81**	2.08	**0.79**	2.08	**0.82**	2.07	**0.75**

TABLE 5.2 Storage Space Requirement for 240 Original and Compressed Images

BPP (Compression Ratio)	File Sizes in Kilo Bytes					
	256 × 256 (80 Images)		512 × 512 (80 Images)		1024 × 1024 (80 Images)	
	Online	Real Time	Online	Real Time	Online	Real Time
Full BPP (1:1)- (original size)	10,485.76	5242.88	41,943.04	20,971.52	167,772.16	83,886.08
0.1 (80:1)	131.07	65.53	524.28	262.14	2097.152	1048.576
0.05 (160:1)	65.54	32.77	262.14	131.07	1048.576	524.288
0.03 (266:1)	39.322	19.66	157.29	78.64	629.150	314.570

transform-based compression techniques. A set of 240 images are used for this analysis. 2D MRI brain, MRI abdomen, CT kidney images, and CT brain images are used to prove the flexibility of the results. The comparison is made in terms of quality measures such as PSNR, SSIM, FSIM, and Q for BPP up to 0.15 (lossy compression). The storage space is calculated before and after compression with respect to bit rates and shown in Table 5.2.

As shown in Table 5.2, when the images are compressed at 0.1 BPP in the place of storing one image, we can store 80 images and increases to 160 and 266 images for 0.05 and 0.03 BPP respectively. The proposed methods are compared with JPEG2000 standards to prove their efficiency. The results are given below for all quality metrics in Fig. 5.8. All the proposed methods perform better than JPEG 2000 in terms of PSNR and FSIM. In case of SSIM alone, PWT performance was more or less equal to JPEG 2000. Considering Q, PCT alone shows less performance than JPEG2000. For all other cases the proposed methods perform better than the standard JPEG2000 algorithm. Comparison is done with other recent techniques also, as shown in Table 5.3. The results were satisfactory in both the comparisons and this proves the superiority of the proposed methods over recent methods from the literature.

The comparative analysis is also done between the proposed methods PWT, PCT, and PRT on different modalities and different sizes of medical images using PSNR, SSIM, FSIM, and Q.

5.3.2.1 Comparative Analysis Based on PSNR and SSIM

PSNR is the ratio of the maximum value of the pixel to the noise (MSE) that affects the quality of the pixels. The higher the PSNR, the smaller the error and it is expressed in terms of a logarithmic decibel scale. The PSNR results

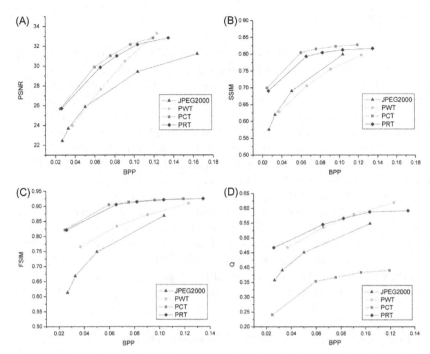

FIGURE 5.8 Comparison between PWT, PCT, and PRT with JPEG2000: (A) PSNR; (B) SSIM; (C) FSIM; and (D) Q.

TABLE 5.3 PSNR Comparison Between the Proposed Methods With References [32,33]

BPP	PSNR (dB)			
	Ref. [32]	PWT	PCT	PRT
2.0	23.53	**60.1**	**37.9**	**37.9**
0.5	22.95	**30.2**	**27.4**	**27.5**
0.125	22.55	**23.1**	22.0	21.1
0.03125	**21.95**	21.0	19.4	19.4
0.0078	**21.94**	15.7	18.5	18.5
BPP	**PSNR (dB)**			
	Ref. [33]	PWT	PRT	PCT
1.25	40.2	**55.7**	38.0	**40.3**
1.0	38.1	**53.2**	37.4	37.6
0.5	33.5	**39.4**	**35.0**	33.3
0.25	30.1	**30.5**	**30.2**	30.7
0.125	26.5	24.0	**26.9**	26.4

for 240 test images are observed for various BPPs and compared between PCT, PRT, and PWT approaches and are graphically shown in Fig. 5.9.

In terms of PSNR, the PWT-based compression approach performs better for CT real-time images for all sizes and for all bit rates. For other modalities, PCT and PRT perform better at lower bit rates and PWT gave good PSNR at higher bit rates. The table shows the BPP at which best PSNR is obtained for all the test images. For 1024 × 1024 sized CT online images, PCT and PRT gave good results compared to PWT at all bit rates. PSNR compares the reference image with the reconstructed image and does numerical comparison, but SSIM takes into account the biological factors of the Human Visual System (HVS) and calculates the structural similarity between them. In case of SSIM, the larger images with size 1024 × 1024 showed better performance using PCT and PRT at all bit rates compared to PWT. For CT real-time images and for 512 × 512 images, PWT showed better performance at higher bit rates. For smaller images with the size 256 × 256, PWT performance is better than others at all the bit rates.

5.3.2.2   Comparative Analysis Based on FSIM and Q

FSIM is another measure which calculates quality based on HVS. SSIM does not vary with the perception of the image, but HVS does. This property is captured in FSIM and a low-level feature comparison between reference image and the distorted image is done to capture this. The results of FSIM for various bit rates are plotted in Fig. 5.10.

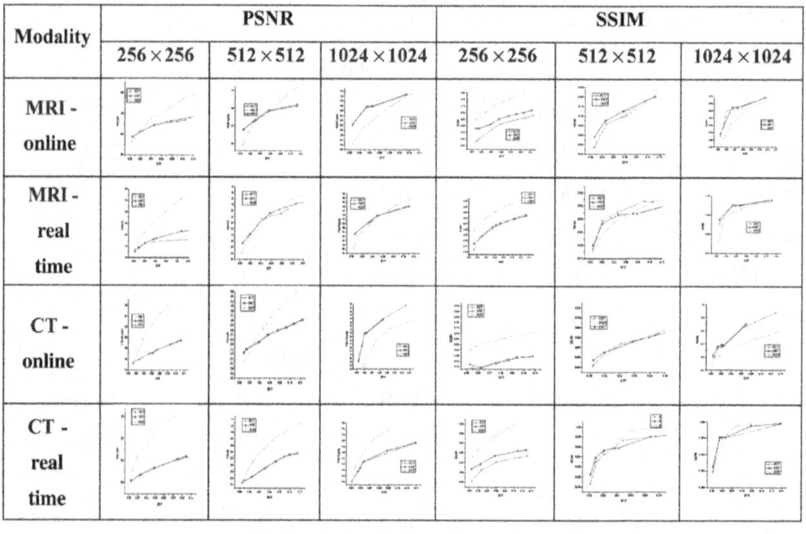

Modality	PSNR			SSIM		
	256 × 256	512 × 512	1024 × 1024	256 × 256	512 × 512	1024 × 1024
MRI - online						
MRI - real time						
CT - online						
CT - real time						

FIGURE 5.9   Comparison of BPP vs. PSNR and BPP vs. SSIM using test images of different sizes and modalities.

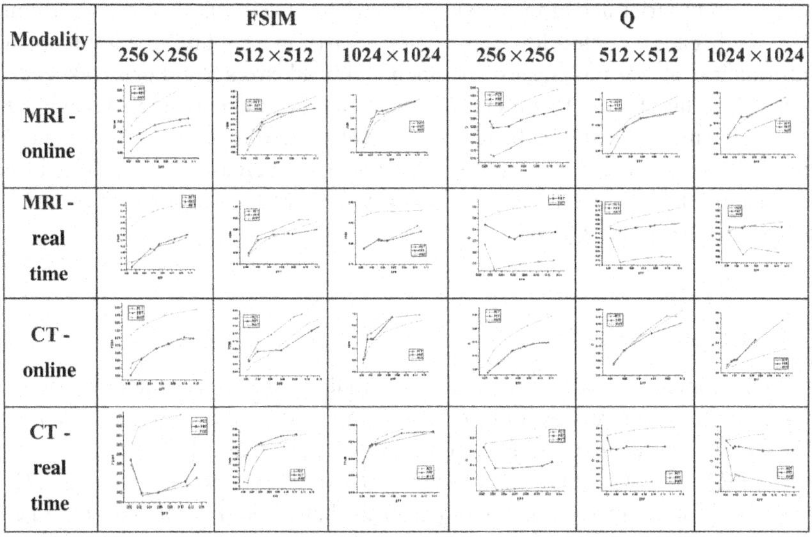

Modality	FSIM			Q		
	256×256	512×512	1024×1024	256×256	512×512	1024×1024
MRI - online						
MRI - real time						
CT - online						
CT - real time						

FIGURE 5.10 Comparison of BPP vs. FSIM and BPP vs. Q using test images of different sizes and modalities.

Considering the FSIM quality metric, the PWT technique proves to perform better than the contourlet- and ripplet-based compression approaches. For CT online images PCT performs better for 512×512 and 1024×1024 sized images. The distortion considered in case of Q is loss of correlation, contrast enhancement, and luminance distortion. Here, HVS is not considered and the values are calculated purely based on pixel values similar to MSE, but the performance is better than MSE in capturing distortion. PWT outperforms all other methods in terms of Q which is the universal quality index. For CT, only online images of size 1024×1024 PCT and PRT performed better.

We have presented several results by applying our proposed wavelet, contourlet, and ripplet transform-based algorithms for compressing real-time and online MRI and CT dataset images of different sizes. These continuous sets of images occupy large storage space and need a lot of time for transmission. A simple preprocessing technique which captures the symmetry between consecutive sets of images and the transform-based compression approach helps in reducing the storage space and time required for transmission. The quality at each bit rate determines which algorithm is better. We have investigated four quality measures which are based on a mathematical comparison approach between the reference image and reconstructed image and HVS to provide results. The wavelet transform-based compression approach performed better for most of the cases. But when it comes to very low bit rates compression, especially for nonregions of interest of medical images, contourlet and ripplet transforms where ahead of wavelet transform. This is because contourlet and ripplet transforms capture the directional information at various scales and their subband sizes are huge; however, with fewer number of bits they give better quality than wavelet. For higher bit rates, smaller sized subbands of wavelet transform helps achieve better quality than the others. As the size of the images increases, the performance of contourlet and ripplet increased more than wavelet, specifically in terms of PSNR, SSIM, and FSIM.

5.4 CONCLUSION

A comparative analysis was done for finding the best-suited transform approach for compressing different modalities of medical images, such as 2D MRI /CT of different sizes 256×256, 512×512, and 1024×1024 using PWT, PRT, and PCT. Quality measures, such as PSNR, SSIM, FSIM, and Q, help in finding which transform performs better at a particular bit rate. Experiments were preformed using a set of 240 images and the results were presented for all the quality measures at various bit rates. The observations are summarized:

1. For 256×256 images: PWT-based compression approach outperforms PCT and PRT for all types of images and for all the quality measures.

   Reason: Higher PSNR, SSIM, FSIM, and Q for PWT are due to the fact that the ringing artifact that occurs in the wavelet based-transformation approach are less prone to the human visual system, while the distortion level is less for smaller images.

2. For 512×512 images: PCT and PRT showed good PSNR, SSIM, FSIM, and Q at lower bit rates and comparatively better performance is obtained using PWT at intermediate or higher bit rates.

   Reason: For intermediate or higher bit rates most of the subband coefficients are captured in case of wavelet transform, which is not the case for other transforms which have larger subbands.

3. For 1024×1024 images: PWT performance was similar to, or higher than, PCT and PRT for MRI/CT real-time images, while for all other images PRT and PCT performed better.

   Reason: For larger images rectangular scaling and the generalized scaling results in reconstructing good quality images compared to small images which have fewer scaling effects.

The results clarify the use of the proposed transform-based compression approaches on different modalities and sizes of medical images for easy transmission and storage. The comparative analysis results clarify that the proposed methods performed better even at lower bit rates and can be used efficiently in the storage and forwarding units of telemedicine applications.

ACKNOWLEDGEMENT

We thank Dr. J. Gnanaraj, Urologist and Director of Medical Services at the Seesha Karunya Community Hospital for giving valuable suggestions and assisting us in our research. His comments helped us to improve the manuscript.

REFERENCES

[1] F. Zhang, Y. Song, W. Cai, A.G. Hauptmann, S. Liu, S. Pujol, et al., Dictionary pruning with visual word significance for medical image retrieval, Neurocomputing 17 (2016) 75–88.

[2] European Society of Radiology, Usability of irreversible image compression in radiological imaging, Insights Imaging 2 (2) (2011) 103–115.

[3] Y. Chen, Medical image compression using DCT-based subband decomposition and modified SPIHT data organization, Int. J. Med. Inform. 76 (2007) 717–725.

[4] B.J. Erickson, Irreversible compression of medical images, J. Dig. Imaging 15 (1) (2002) 5–14.

[5] C. Doukas, I. Maglogiannis, Region of interest coding techniques for medical image compression, IEEE Eng. Med. Biol. Mag. 26 (5) (2007) 29–35.

[6] J. Bartrina-Rapesta, J. Serra-Sagristà, F. Aulí-Llinàs, JPEG2000 ROI coding through component priority for digital mammography, Comput. Vis. Image Underst. 115 (1) (2011) 59–68.

[7] P. Akhtar, M. Bhatti, T. Ali, M. Muqeet, Significance of roi coding using maxshift scaling applied on MRI images in teleradiology-telemedicine, J. Biomed. Sci. Eng. 1 (02) (2008) 110–115.

[8] G. Vallathan, K. Jayanthi, Region based image compression based on textural properties in MRI images for telemedicine applications, Dig. Image Process. 5 (7) (2013) 357–362.

[9] I. Maglogiannis, C. Doukas, G. Kormentzas, T. Pliakas, Wavelet-based compression with ROI coding support for mobile access to DICOM images over heterogeneous radio networks, IEEE Trans. Inform. Technol. Biomed. 13 (4) (2009) 458–466.

[10] G.M. Francois, R.C. Ronald, Brushlets: a tool for directional image analysis and image compression, Appl. Comput. Harmon. Anal. 4 (1997) 147–187.

[11] E.L. Pennec, S. Mallat, Sparse geometric image representations with bandelets, IEEE Trans. Image Process. 14 (4) (2005) 423–438.

[12] V. Velisavljevic, B. Beferull-Lozano, M. Vetterli, P.L. Dragotti, Directionlets: anisotropic multi-directional representation with separable filtering, IEEE Trans. Image Process. 15 (7) (2006) 1916–1933.

[13] P.R. Deshmukh Miete, A.A. Ghatol Fiete, Multiwavelet and image compression, IETE J. Res. 48 (3–4) (2002) 217–220.

[14] M.N. Do, M. Vetterli, The contourlet transform: an efficient directional multiresolution image representation, IEEE Trans. Image Process. 14 (12) (2005) 2091–2106.

[15] J. Xu, L. Yang, D. Wu, Ripplet: a new transform for image processing, J. Vis. Commun. Image Represent. 21 (7) (2010) 627–639.

[16] S.S. Tamboli, V.R. Udupi, Image compression using haar wavelet transform, Int. J. Adv. Res. Comput. Commun. Eng. 2 (8) (2013).

[17] B. Andries, J. Lemeire, A. Munteanu, Scalable texture compression using the wavelet transform, Vis. Comput. (2016). Available from: https://doi.org/10.1007/s00371-016-1269-1.

[18] Ali Hassan, H.R. Raghad, Gabor wavelet transform in image compression, J. Kufa Math. Comput. 1 (6) (2012) 107–113.

[19] S.M. Hosseini, A.R. Naqhsh-Nilchi, Medical ultrasound image compression using contextual vectorquantization, Comput. Biol. Med. 42 (7) (2012) 743–750.

[20] A.K.J. Saudagar, A.S. Syed, Image compression approach with ridgelet transformation using modified neuro modeling for biomedical images, Neural Comput. Appl. 24 (7) (2014) 1725–1734.

[21] V.K. Bairagi, Symmetry-based biomedical image compression, J. Digit. Imaging 28 (6) (2015) 718–726.

[22] N.B. Erichson, S. Voronin, S.L. Brunton, J.N. Kutz, arXiv. 1608.02148 Randomized Matrix Decompositions Using R, arXiv preprint, 2016.

[23] I. Baeza, J.A. Verdoy, J. Villanueva-Oller, R.J. Villanueva, ROI-based procedures for progressive transmission of digital images: a comparison, Math. Comput. Model. 50 (5) (2009) 849–859.

[24] A.M. Rufai, G. Anbarjafari, H. Demirel, Lossy image compression using singular value decomposition and wavelet difference reduction, Digit. Signal Process. 24 (2014) 117–123.

[25] F. Kamisli, A low-complexity image compression approach with single spatial prediction mode and transform, Signal Image Video Process. 10 (2016) 1409–1416.

[26] D. Zhao, S. Zhu, F. Wang, Lossy hyperspectral image compression based on intra-band prediction and inter-band fractal encoding, Comput. Electr. Eng. 54 (2016) 494–505.

[27] D. Venugopal, S. Mohan, S. Raja, An efficient block based lossless compression of medical images, Optik 127 (2016) 754–758.

[28] R. Herrero, V.K. Ingle, Ultraspectral image compression using two-stage prediction: prediction gain and rate-distortion analysis, Signal Image Video Process. 10 (2016) 729–736.

[29] P. Eben Sophia, J. Anitha, Contextual MRI image compression using normalized Wavelet-transform coefficients and prediction, IETE J. Res. 63 (5) (2017) 671–683.

[30] P. Eben Sophia, J. Anitha, Contourlet transform based sub-band normalization for region based medical image compression, Intell. Decis. Technol. 10 (4) (2016) 385–391.

[31] P. Eben Sophia, Contextual Medical Image Compression Algorithms Using Image Transforms and Optimized Prediction Model, Doctoral Thesis, Karunya Institute of Technology and Sciences, May 2017.

[32] H. Abir Jaafar, Al. J. Dhiya, R. Naeem, P.J.G. Lisboa, Hybrid neural network predictive-wavelet image compression system, Neurocomputing 151 (3) (2014) 975–984.

[33] K. Ahmadi, A.Y. Javaid, E. Salari, An efficient compression scheme based on adaptive thresholding in wavelet domain using particle swarm optimization, Signal Process. Image Commun. 32 (2015) 33–39.

Chapter 6

A Medical Image Retrieval System in PACS Environment for Clinical Decision Making

Vijay Jeyakumar[1] and Bommannaraja Kanagaraj[2]

[1]*Department of Biomedical Engineering, SSN College of Engineering, Chennai, India,*
[2]*Department of Electronics and Communication Engineering, KPR Institute of Engineering and Technology, Coimbatore, India*

Chapter Outline

6.1 INTRODUCTION

Currently, an enormous amount of medical data is produced day by day in healthcare sectors to diagnose the disease. These images are used by healthcare professionals in various applications, such as critical care diagnosis, Tele−surgeries, case-based reasoning, Tele−education, and medical research. The integration of Picture Archives and Communication System (PACS), Electronic Patient Record (EPR), and Hospital Information system (HIS) (illustrated in Fig. 6.1) in any healthcare sector results in filmless and paperless health reports, increase work progress, reduced operation costs, and improved storage of voluminous data. Many methods and algorithms are proposed by various professionals to retrieve the stored data for the past six decades [1].

Every day, billions of data are searched for by different kinds of internet users; many searches for, and find, images of interest. These images are varied

Intelligent Data Analysis for Biomedical Applications. DOI: https://doi.org/10.1016/B978-0-12-815553-0.00006-9

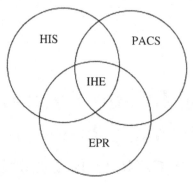

FIGURE 6.1 Integrated healthcare environment.

in visual and semantic content, location, size, and how they originate. Image searches can be classified as search by association, search by category, and aimed search. Any image retrieval system can be visualized from both the user's perspective and that of the system. Smeulders et al. [2] categorize users as browsers, surfers, and searchers, based on the user's intention for their search. Another important factor involved in an image retrieval system is the type of query. From the system's perspective, various query modalities available are keywords, images, graphics, and a composite of these. A common image format, DICOM (Digital Imaging and Communication in Medicine), is followed by healthcare entities to integrate imaging modalities of various kinds. Image representation provides strong evidence in most kinds of disease diagnoses and case-based reasoning. is The major task in Radiological Information System (RIS) is to access the right information from the huge repository at the time of the search and delivery to the right people. In the medical domain, various health standards, such as Unified Medical Language System (UMLS), Health Level Seven (HL7), Americal Standards for Testing and Materials (ASTM), and Systemized Nomenclature of Medicine (SNOMED) were proposed by professionals to achieve interoperability. A similar standard framework has not yet been proposed to retrieve images from the PACS environment. Despite this drawback, some common approaches are followed by hospital personnel to retrieve the desired medical images from a database. They are: (1) Text-Based Image Retrieval (TBIR) [3]; (2) Content-Based Image Retrieval (CBIR) [2]; and (3) Semantic-Based Image Retrieval (SBIR).

6.1.1 Text-Based Image Retrieval System

The text-based retrieval approach is a traditional method followed by professionals in hospitals and for various commercial applications, such as Google Image searches, Flickr, and YouTube. Limited keywords, tags, patient IDs, and other textual attributes related to the report are used to retrieve medical images from the PACS environment [4]. Since the TBIR system is based on language, dissimilarities in annotations of the images may increase the system's complexity, creating a greater impact on human effort, cost-effectiveness, and timeliness. As TBIR depends on the subjectivity of human perception about the input query and results, irrelevant objects may be suggested by the system, which ultimately degrades the system's performance. Although the TBIR system is fast and reliable, it becomes incapable if the images are non-annotated.

6.1.2   Content-Based Image Retrieval System

To overcome the issues pertaining to TBIR, like being time-consuming and demanding more human effort, the content-based (or feature-based) image retrieval system was proposed. The extraction of visual features from the huge repository is possible through minimal human intervention and with reduced errors. Some of the commercially available applications based on CBIR are Query by Image Content (QBIC) by Flickner et al. [5], Virage's VIR Image Engine [6], Visualseek [7], Flexible Image Retrieval Engine (FIRE), RetrievalWare [8], imgseek, and the GNU image finding tool. These systems adopt many image processing techniques to extract local and global features, such as the color, shape, texture, energy, and various statistical features from the image [9]. This automated paradigm is more powerful and can handle a huge number of medical images in the PACS environment. Another important factor of adopting CBIR in the healthcare environment is its potential to reduce the gap between human perception and semantic concept. However, problems still exist due to inadequate knowledge about the characterization of images. Radiographs are characterized by origin, anatomical structure, orientation, and pathophysiological information, among others. In CBIR, the images are indexed and annotated by its various features and morphological appearances only. A common CBIR system does not provide meaningful information to radiologists about one or more objects present in an image, abnormal findings, and the most significant features for case-based diagnosis. Hence, using an effective CBIR system to find relevant images from the repository remains a challenging task.

6.1.3   Semantic-Based Image Retrieval System

SBIR focuses on image information such as anatomy, modalities, and other pathological information. MedGIFT, CasImage, Yottalook, iMedline, Flexible Image Retrieval Engine (FIRE), Spine pathology and Image retrieval system (SPIRS), and Yale Image Finder (YIF) were proposed for various case studies and conditions. However, the image processing techniques and text annotations adopted by each system differ.

Medical professionals' expectation from any found image should include related information, such as patient information, medical examination procedures and reports, symptoms, interpretation of radiologists, the timeline of an image captured, directionality, and body region examined, etc. Extraction of features from images does not yield meaningful results. Both TBIR and CBIR have limitations. By considering the advantages of both the methods, a new retrieval framework can be formulated based on users' intent to increase the efficiency of the overall system. A hybrid retrieval mechanism enables professionals to use text and/or image-based queries in the system.

6.2 PROPOSED INTEGRATED FRAMEWORK

The main purpose of an integrated framework is to retrieve heterogeneous medical images from the PACS environment effectively, to provide high-quality health care, especially during critical care conditions. The proposed framework is constructed by three phases, namely CBIR, TBIR, and their Fusion/Hybrid (a combination of CBIR and TBIR). In each phase, the retrieval performance is evaluated and compared. The overall system framework is provided in Fig. 6.2.

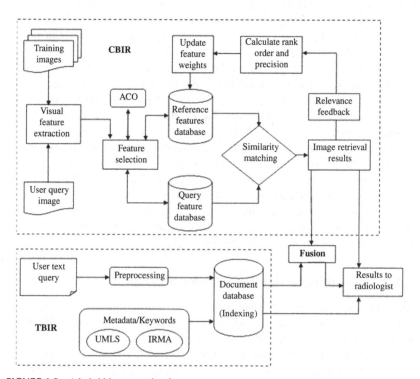

FIGURE 6.2 A hybrid image retrieval system.

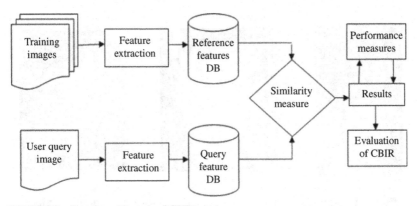

FIGURE 6.3   Generic architecture of CBIR system.

In CBIR systems, the significant features of testing and training images are extracted and the similarity between them is calculated. Likewise, the input text query is compared with image annotations and its similarity indices are calculated. This kind of approach is called the Query by Example (QBE) paradigm. For this proposed framework, a database consists of 12,076 training images and 1000 testing images with the representations of age, gender, orientation, and the specifications of the acquired modality are obtained from University of Aachen, Germany, by Deserno and Ott [10]. All images $I(x,y)$ are scaled down to 512×512 due to their diversified aspect ratio. Also, all the images in the database are labeled and classified by medical professionals based on the IRMA (Image Retrieval in Medical Applications) code. These images reflect various anatomical regions, such as the spine, breast, upper and lower extremities, skull, and chest.

6.2.1   Phase I: Content-Based Image Retrieval

To meet the objectives of the proposed method, heterogeneous images of various kinds are considered and the features of both training and testing images are extracted. A simple architecture of CBIR is initially undertaken to perform the task, as illustrated in Fig. 6.3.

For an experimental analysis, eight query images are randomly selected without any classification from the testing images. These images are given in Fig. 6.4 with their descriptions. The entire training image database with classifications and annotations is considered for the proposed framework of the CBIR and TBIR segment. This retrieval framework is developed by MATLAB® software.

6.2.1.1   Feature Extraction

The preliminary stage of the CBIR system is to extract the features of both training $P(x,y)$ and testing/query images $Q(x,y)$. Based on the literature

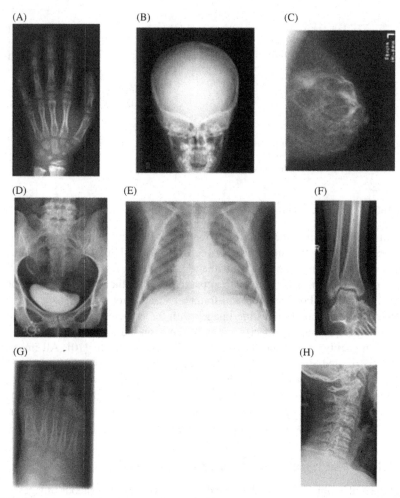

FIGURE 6.4 Query images. (A) Query 1: X-ray–analog overview–right anterior obli-que–upper extremity (hand)–musculoskeletal system. (B) Query 2: X-ray–analog-over-view–occipitomental–nose area–musculoskeletal system. (C) Query 3: X-ray–analog–low beam energy–oblique–breast (mamma)–female system. (D) Query 4: X-ray–analog overview image–unspecified pelvic. (E) Query 5: X-ray–analog overview–supine–lower middle quad-rant–uropoietic system. (F) Query 6: X-ray–analog–high energy beam–posteroanterior–chest. (G) Query 7: X-ray–analog overview–reclination–lower extremity–musculoskeletal system. (H) Query 8: X-ray–analog overview–cranium–cervical spine–musculoskeletal system. *Courtesy TM Deserno et al.*

study, significant features of all medical images are extracted. To ensure the human perception about the images, coarseness, contrast, and directionality is calculated [11]. Texture information and frequency components of the images are other important elements to be considered during the feature

TABLE 6.1 Number of Features Extracted

Feature Descriptors	Tamura	Gabor	Region Based	Wavelet Moments	Global Textures	HOG	Moment Invariants	Total Feature Vectors
Number of features	03	24	08	64	12	81	07	199

extraction process. To retain the spatial information, Gabor features are determined by convolving the windowed function with $I(x, y)$. As each image contains more anatomical representation, region-based information extraction is the most appropriate [12]. To attain this, Discrete Cosine Transform (DCT), major and minor axis length, eccentricity, convex area, filled area, equivalence diameter, and contrast of each image are calculated. To determine the texture information and intensity level distributions in images, energy, entropy, inertia, gray level co-occurrence matrix (GLCM) [13], and the histogram of orientation descriptor [14] are applied to express the contour and manifestation of objects in the images. Due to certain variations in position, size, orientation, and reflection in images, determining the moment values of each dataset is noteworthy. The number of features extracted using various methods is listed in Table 6.1.

6.2.1.2   Similarity Measures

The similarity between the features of the query and training images are evaluated by the distance measurement techniques listed in Table 6.2.

From the list of distance measures, the closeness of the query and training feature descriptors are calculated. The most associated features representing the images are retrieved by the system for the user. From Figs. 6.5 and 6.6, it is evident that the retrieved images do not represent the inputs of query 2 and 8. It shows that the feature values of irrelevant retrieved images are closer to the input image features. Another important inference observed from these figures is that a few feature descriptors of the retrieved images may not be significant for the image content of the query image. These insignificant features suggest that irrelevant images will appear.

Neither a standard testing database nor measurement method is available; the efficiency of the framework is measured by two important metrics:

Precision: The ratio of retrieved documents that are most related to the users' intent. In this framework, the total number of retrieved documents to be displayed for the review is 20.

TABLE 6.2 Types of Similarity Measures

Euclidean	$d_{Euc} = \sqrt{\sum_{i=1}^{d} \lvert p_i - q_i \rvert^2}$
	Where
	p_i is the reference image descriptor values
	q_i is the query image descriptor values
	d is the total number of features extracted
Manhattan	$d_{mh} = \sum_{i=1}^{d} \lvert p_i - q_i \rvert$
Relative deviation	$d_{rd} = \dfrac{\sum_{i=1}^{d} \lvert p_i - q_i \rvert^2}{\frac{1}{2}\left(\sum_{i=1}^{d} q_i^2\right) + \left(\sum_{i=1}^{d} p_i^2\right)}$
Mahalanobis	$d_{mb} = \sqrt{\sum_{i=1}^{d} \dfrac{\lvert p_i - q_i \rvert^2}{s_i}}$
	S_i is the standard deviation of pi and qi.
Cosine distance	$d_{cd} = \dfrac{\sum_{i=1}^{d} p_i q_i}{\frac{1}{2}\left(\sum_{i=1}^{d} q_i^2\right) + \left(\sum_{i=1}^{d} p_i^2\right)}$
Chebyshev	$d_{cheb} = \max\left(\lvert p_i - q_i \rvert\right)_{i=1,2,\dots d}$
Spearman rank	$d_{spr} = \dfrac{6\sum_{i=1}^{d} (\text{rank}(p_i) - \text{rank}(q_i))^2}{d(d^2 - 1)}$

$$\text{Precision} = \frac{\lvert \{\text{Relevant documents}\} \cap \{\text{Retrieved documents}\} \rvert}{\lvert \{\text{Retrieved documents}\} \rvert} \quad (6.1)$$

Recall: The ratio of retrieved documents, which are related to users' intent to related documents available in the database.

$$\text{Recall} = \frac{\lvert \{\text{Relevant documents}\} \cap \{\text{Retrieved documents}\} \rvert}{\lvert \{\text{Relevant documents}\} \rvert} \quad (6.2)$$

For various input images of Queries 1–8, the precision and recall values for different similarity measures are listed in the Table 6.3.

The highlighted values in Table 6.3 resembles that Query 3 and 5 inputs to the CBIR system yields the highest precision value of 1 for Euclidean, Mahalanobis, and Spearman distance measurement methods. For the same input query images, other similarity measures provide low precision values.

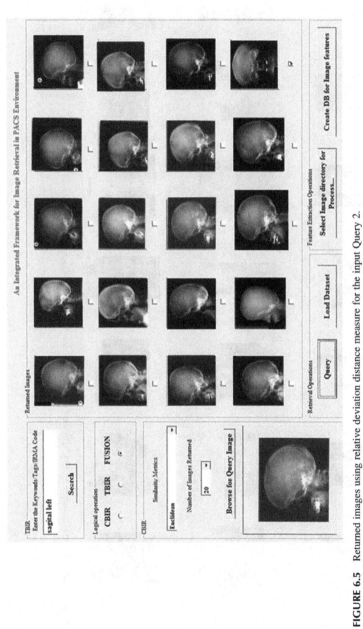

FIGURE 6.5 Returned images using relative deviation distance measure for the input Query 2.

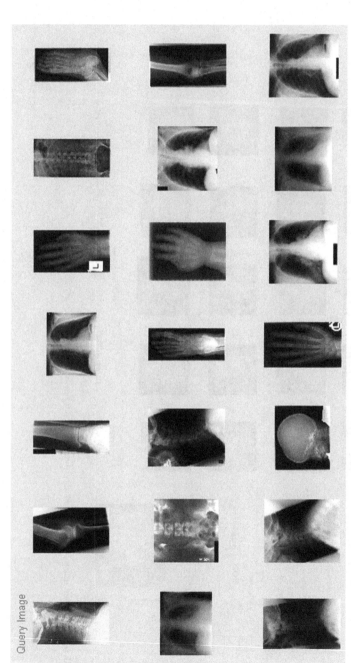

FIGURE 6.6 Returned images using Chebyshev distance measure for the input Query 8.

TABLE 6.3 Similarity Measures between Query and Reference Images Using Generic CBIR Architecture

Query Images	Euclidean		City Block		Relative Deviation		Mahalanobis		Cosine		Chebyshev		Spearman		MAP
	P	R	P	R	P	R	P	R	P	R	P	R	P	R	
1	0.25	0.02	0.20	0.01	0.15	0.01	0.40	0.03	0.25	0.02	0.20	0.01	0.10	0.01	0.22
2	0.15	0.03	0.05	0.01	0.05	0.01	0.05	0.01	0.00	0.00	0.10	0.02	0.00	0.00	0.06
3	1.00	0.08	0.85	0.07	0.45	0.04	0.50	0.04	0.30	0.03	0.35	0.03	1.00	0.08	0.64
4	0.55	0.13	0.15	0.04	0.10	0.02	0.05	0.01	0.15	0.04	0.10	0.02	0.40	0.09	0.21
5	1.00	0.04	0.85	0.04	0.85	0.04	1.00	0.04	0.65	0.03	0.80	0.04	1.00	0.04	0.46
6	0.45	0.06	0.05	0.01	0.15	0.02	0.15	0.02	0.15	0.02	0.25	0.03	0.20	0.03	0.20
7	0.35	0.07	0.00	0.00	0.00	0.00	0.05	0.01	0.10	0.02	0.00	0.00	0.25	0.05	0.11
8	0.70	0.26	0.05	0.02	0.10	0.04	0.00	0.00	0.00	0.00	0.15	0.06	0.35	0.13	0.19

P—precision; R—recall; MAP—mean average precision.

The precision and recall values "0" indicate that no related images from the database are retrieved. The retrieval system's efficiency can be derived from the mean average value of precision pertaining to individual query images. The overall system efficiency is achieved at a rate of less than 40%. Again, these varied values of the same distance measurement technique for different input queries or the same input query for different distance measures resembles the inconsistency between the feature descriptors. In order to achieve consistency among the distance measures and features, the selection of the best features/sub-sets is essential. This can be achieved by adopting suitable optimization techniques during the feature selection.

6.2.1.3 Feature Selection

Feature selection is another important and commonly used technique in data processing to select appropriate features from noisy data. This kind of approach increases the execution speed of any data processing algorithm that increases the accuracy of prediction and reduces the variable results [15]. Feature selection in medical data processing is unavoidable due to its increased performance and reduces computational cost. To identify suitable features from the extracted features of training and testing images, WEKA software [16] is used to visualize significant features. Based on the functions available, Correlation-based Feature Selection (CFS) and ranker approaches are chosen. Under CFS, Linear Forward Selection (LFS) [17] and Genetic Algorithm (GA) are selected. Under the ranker category, Principle Component Analysis (PCA) and GA are selected to find the best sub-sets. All feature selection testing is performed based on 5-fold and 10-fold cross-validation. These results do not provide huge variations in the results. From the results obtained, the number of best sub-sets suggested by each method is listed in Table 6.4.

As per the number of features suggested by each method, the best sub-sets of both testing and training images are only considered and their similarity is calculated for the input of Query images 1−8. Among precision

TABLE 6.4 Best Sub-Sets Suggested by CFS and Ranker Approaches

Approach	Search Method	Number of Suggested Sub-Sets
Correlation based feature selection (CFS)	Linear forward selection (LFS)	105
Ranker	Genetic algorithm (GA)	88
Ranker	Principle component analysis (PCA)	148
CFS	Genetic algorithm	88

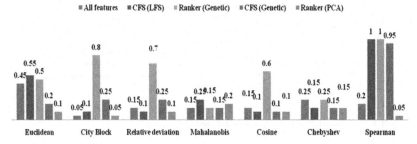

FIGURE 6.7   Comparative precision values for various similarity measures of Query 6.

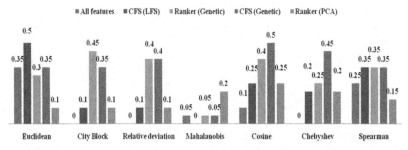

FIGURE 6.8   Comparative precision values for various similarity measures of Query 7.

and recall, precision plays a vital role to calculate retrieval efficiency. Hence, hereafter all data results are listed and plotted based on the precision values. The comparative analysis performance of all feature sets and best sub-sets for the input Query images of 6 and 7 are shown in Figs. 6.7 and 6.8.

From the results obtained, the inconsistency among the features remains the same by comparing the precision values of all the similarity measures. This happens due to the functionality of LFS and the Ranker-based feature selection that follow the wrapper approach. In the wrapper approach, the learning algorithm suggests the best features. The performance of the wrapper approach does not reflect in the selection of best sub-sets. In addition, the overall retrieval performance has been raised to only 45%, which is not acceptable in the medical domain. In order to overcome this problem, an intelligent feature selection comprising a learning algorithm and feedback mechanism can be applied. This type of feature selection, called the embedded approach, is illustrated in Fig. 6.9.

Recently, embedded-based feature selection pays more attention to data science researchers to get optimistic best sub-sets on real-time data. The embedded feature selection approach searches and iterates until the best sub-set S_0 at increased cardinality C is found. At each cardinality, the present sub-set is compared with the previous best sub-set. To learn the best sub-sets

Selecting the best subset

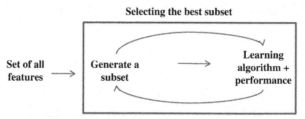

FIGURE 6.9   Block diagram of the embedded approach.

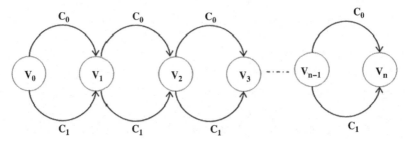

FIGURE 6.10   The digraph.

at each iteration, Ant Colony Optimization (ACO) and relevance feedback are incorporated into the framework.

6.2.1.4   Ant Colony Optimization Method

Dorigo M. proposed a new paradigm to design heuristic algorithms based on the social behavior of ants. ACO is popular for various applications, such as the traveling salesman problem, vehicle routing algorithms, data communication problems, speech recognition, and weather forecasting, etc. Although ants are blind, communication between ants is very interesting by their secretion of a chemical from the abdomen called pheromones. During the search for food search, the ants follow one another, one by one, by sensing the pheromones laid by each ant along the path. This mechanism is applied for any kind of optimization technique by considering the factors, such as graph representation, heuristic desirability, feedback process, constraint satisfaction, and solution arrival method/stopping criteria. Chen et al. [18] have proposed a new method of ACO to select the best features for image analysis with O(n) arcs. From Fig. 6.10, the nodes (Vo to Vn) are considered as features and graph-containing connecting arcs are C_j^o and C_j^1 with adjacent nodes v_{j-1} and v_j.

For every arc C_i^j, artificial pheromone value τ_i^j is assigned as the feedback value to direct the ants searching on the graph. Initial pheromone value is $\tau_i^j = 1$ for all $i = 1,2,\ldots,n$ and $j = 0,1$. The probability of an ant to select the path C_i^j to reach v_i from v_{i-1} is determined by pheromone density and heuristic information. This transition rule is calculated by:

$$p_i^j(t) = \frac{\left[\tau_i^j(t)\right]^\alpha \left(\eta_i^j\right)^\beta}{\left[\tau_i^0(t)\right]^\alpha \left(\eta_i^0\right)^\beta + \left[\tau_i^1(t)\right]^\alpha \left(\eta_i^1\right)^\beta} ; (i = 1, 2, \ldots, n \text{ and } j = 0, 1) \quad (6.3)$$

where

1. $\tau_i^j(t)$ is the pheromone on the arc C_i^j.
2. η_i^j is the heuristic information revealing the desirability to select arc C_i^j.
3. α & β are the parameters yield the relevance of pheromone and heuristic.
   a. If, $\alpha = 0$ no feedback information due to loss of previous search experience.
   b. $\beta = 0$ arc is neglected and search becomes entirely random search.

The value of η_i^1 $(i = 1,..,n)$ is calculated from the precision value of similarity measures using the weighted relevance feedback method. As η_i^1 is the denominator of p_i^j, the probability of an ant to select the subsequent path is inversely proportional to the precision value (from the weighted relevance feedback) and the value of $\eta_i^0 = \frac{\xi}{n}\sum_{i=1}^{n}\eta_i^1$, where $\xi \in (0,1)$ is constant.

6.2.1.4.1 Updating the Pheromone

In every iteration, the proposed algorithm updates the pheromone based on the following equations,

$$\tau_i^j(t) = \rho\tau_i^j(t) + \Delta\tau_i^j(t) + Q_i^j(t) \quad (6.4)$$

where

$$\Delta\tau_i^j(t) = \frac{1}{|S_i^j|}\sum_{s\in S_i^j}f(s) \quad (6.5)$$

and

$$Q_i^j(t) = \begin{cases} Q, & C_i^j \in S_{\text{best}} \\ 0, & \text{otherwise} \end{cases} \quad (6.6)$$

S_i^j is the best solution presented at ith iteration through C_i^j. S_{best} is the best sub-set and Q is positive constant.

6.2.1.5 Feature Fusion

To combine the features of the best sub-set, the late fusion method is proposed due to its characteristic of treating all the features equally for various image queries. The initial weights are assigned to all the features subject to $0 \le \omega^F \le 1$ and $\sum \omega^F = 1$.

The similarity between the query image, qi, and training images, pi, are represented by:

$$\text{Sim}(q_i, p_i) = \sum_F \omega^F \text{Sim}^F(q_i, p_i) \tag{6.7}$$

where $F \in$ are all the extracted features and ω^F is the weight of different image representations.

6.2.1.6   Weighted Relevance Feedback

The modified feedback system comprises weight adjustment parameters, efficiency rating (E), and the calculation of rank order (K) and precision (P). The entire process is formulated as an algorithm which is illustrated in Table 6.5. To select the best features and update the weights of the features, the P value of each similarity matching is substituted in Eq. (6.3) to obtain the subsequent optimal feature subset.

In an ACO algorithm, the initial pheromone intensity of an arc and heuristic value is fixed at 1. The total number of nodes in the digraph is 199. The initial value of η_i^1 is zero. The derived precision and recall values using ACO and WRF are listed in Table 6.6 for various input query images.

TABLE 6.5   Weighted Relevance Feedback Algorithm

Initialize equal weights to all features
Find initial similarity matching between qi and pi
{
    Relevance feedback starts
    {
    Update
    {
    Find the efficiency of returned images

$E = \dfrac{\sum_{i=1}^{k} \text{Rank(i)}}{k/2} . P(k)$ // effectiveness of top k returned images

    Where Rank (i) = 0; for irrelevant images in i position
    Rank (i) = (k-i) / (k-1); for relevant images
    Determine the precision value
    P(k) = R_k/k; // Precision at top k
    Where
    R_k the number of related images in top K retrieved results
    }
    Update feature weights by $É = \omega F$.
    Find similarity matching between qi and pi by
    $\text{Sim}(q_i, p_i) = \sum_F \omega^F \text{Sim}^F(q_i, p_i)$
    }
    Determine top K returned images
}
Stop the retrieval process when user gets relevant images.

TABLE 6.6 Performance Measures of the Retrieval System Using ACO and Relevance Feedback

Query	Euclidean		City Block		Relative Deviation		Mahalanobis		Cosine		Chebyshev		Spearman		MAP
	P	R	P	R	P	R	P	R	P	R	P	R	P	R	
1	0.90	0.06	0.85	0.06	0.85	0.06	0.90	0.06	0.90	0.06	0.90	0.06	0.90	0.06	**0.89**
2	0.85	0.06	0.80	0.05	0.80	0.05	0.75	0.05	0.75	0.05	0.75	0.05	0.75	0.05	**0.82**
3	1.00	0.08	1.00	0.08	0.95	0.08	0.95	0.08	0.95	0.08	0.95	0.08	0.95	0.08	**0.90**
4	0.90	0.21	0.85	0.20	0.85	0.20	0.80	0.19	0.85	0.20	0.80	0.19	0.90	0.21	**0.86**
5	0.95	0.04	0.80	0.03	0.80	0.03	0.80	0.03	0.80	0.03	0.95	0.04	0.95	0.04	**0.86**
6	0.85	0.11	0.85	0.11	0.85	0.11	0.75	0.09	0.80	0.10	0.80	0.10	0.85	0.11	**0.82**
7	0.90	0.19	0.80	0.16	0.80	0.16	0.80	0.16	0.85	0.18	0.85	0.18	0.85	0.18	**0.84**
8	0.85	0.31	0.85	0.31	0.90	0.33	0.90	0.33	0.9	0.33	0.90	0.33	0.80	0.30	**0.87**

From Table 6.6, it is concluded that the maximum precision value is attained by choosing Euclidean distance for the input Query image 3. The overall CBIR system performance results in 87% by adopting ACO and weighted relevance feedback mechanisms. The precision and recall values for the input Query image 4 is compared with the results of other features selection methods. These results are given in Table 6.7.

The returned images for the Query image of 4 using Euclidean distance measurement method is shown in Fig. 6.11.

The ACO technique and weighted relevance feedback mechanism has experimented with all Query images 1−8. Their performance based on different feature selection methods for various values of K returned images shown in Figs. 6.12 and 6.13.

6.3 PHASE II: TEXT-BASED IMAGE RETRIEVAL SYSTEM

As stated in the proposed framework given in Fig. 6.3, the text annotations of the images are extracted and compared with publicly available databases such as UMLS and IRMA. The IRMA database has a symbolic representation of medical images describing its modality technique, directionality, anatomical, and biological information. These are generally represented as TTTT-DDD-AAA-BBB. The complete notion of the IRMA code is available in [19]. In TBIR, the categorization of terms in the documents is performed by assigning weights to the terms. The weight of the term is calculated as:

$$w(t, d) = \frac{g(t, d)}{(1 - \lambda).c + \lambda.n_1(d)} \quad (6.8)$$

$n_1(d)$ is the number of a word's occurrence only once (singleton) and c represents the average number of singletons in all documents, $\lambda \in \{0 \to 1\}$ and $g(t,d)$ is defined as:

$$g(t, d) = \begin{cases} \dfrac{[1 + \log n(t, d)]}{[1 + \log \bar{n}(d)]}, & \text{if } t \in d \\ 0 & , \text{ else} \end{cases} \quad (6.9)$$

$n(t,d)$ is the occurrence of a term's frequently in the document d and $\bar{n}(d)$ is average term frequency in d. To provide weights to the terms in query documents $w(t,q)$ is calculated as:

$$w(t, q) = [1 + \log n(t, q)].idf(t) \quad (6.10)$$

Inverse document frequency (IDF) is calculated as:

$$idf(t) = \log \left[\frac{k}{n(t)} \right] \quad (6.11)$$

TABLE 6.7 Comparative Results of Various Feature Selection Methods for Query 4

Feature Selection Methods	Euclidean		City Block		Relative Deviation		Mahalanobis		Cosine		Chebyshev		Spearman	
	P	R	P	R	P	R	P	R	P	R	P	R	P	R
All features	0.55	0.13	0.15	0.04	0.10	0.02	0.05	0.01	0.15	0.04	0.05	0.01	0.15	0.04
Ranker (GA)	0.25	0.06	0.10	0.02	0.10	0.02	0.20	0.05	0.15	0.04	0.05	0.01	0.10	0.02
CFS (LFS)	0.15	0.04	0.25	0.06	0.15	0.04	0.05	0.01	0.20	0.05	0.05	0.01	0.30	0.07
CFS (GA)	0.25	0.06	0.10	0.02	0.20	0.05	0.15	0.04	0.15	0.04	0.15	0.04	0.55	0.13
Ranker (PCA)	0.30	0.07	0.15	0.04	0.10	0.02	0.05	0.01	0.15	0.04	0.05	0.01	0.15	0.04
ACO + RF	**0.90**	**0.21**	**0.85**	**0.20**	**0.85**	**0.20**	**0.80**	**0.19**	**0.85**	**0.20**	**0.80**	**0.19**	**0.90**	**0.21**

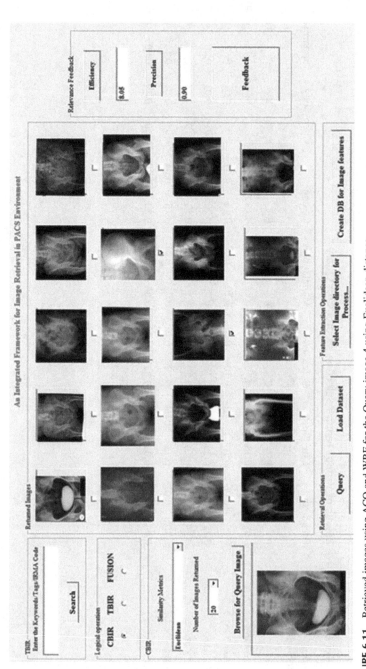

FIGURE 6.11 Retrieved images using ACO and WRF for the Query image 4 using Euclidean distance.

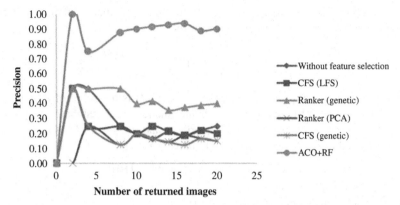

FIGURE 6.12 Performance analyses of various feature selection methods for Query 1 image using Euclidean.

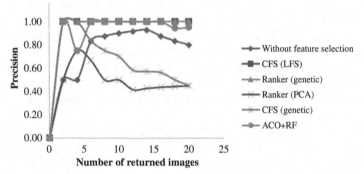

FIGURE 6.13 Performance analyses of various feature selection methods for Query 5 image using Chebyshev measure.

where k is the number of documents and $n(t)$ is the number of documents in which t occurs. The calculated ranking value, retrieval status value (RSV), is the relevance of a document d for a query q is:

$$\text{RSV}(q, d) = \sum_{t \in T} w(t, q).w(t, d) \tag{6.12}$$

T is the set of all the documents.

6.3.1 Phase III: Fusion Retrieval

The fusion/mixed retrieval is attained by considering the advantages of CBIR and TBIR. The results obtained from visual and textual information are fused together in Eq. (6.13). The similarity between a mixed query

$Q = (Q_I, Q_T)$ $\{Q_I \rightarrow$ image(s); $Q_T \rightarrow$ text$\}$ and a medical document (t,d) annotated with medical images is given by:

$$\lambda(Q, I, d) = \alpha \frac{\lambda_v(Q_I, I)}{\max_{z \in D_I} \lambda_v(Q_I, I)} + (1 - \alpha)\frac{\lambda_T(Q_T, d)}{\max_{z \in D_T} \lambda_T(Q_T, z)} \qquad (6.13)$$

where z is the descriptors in the database, $\lambda_v(Q_I, I)$ is the score of the visual similarity between the query image Q_I and images in the database. $\lambda_T(Q_T, d)$ is the similarity between the textual query Q_T and document. D_I is the image database and D_T is the text database. Usually α varies from 0 to 1.

1. If α is 0, then the proposed framework enables TBIR.
2. If α is 1, CBIR is enabled.
3. If α is 0.5, hybrid retrieval is enabled.

From Figs. 6.14 and 6.15, the text query is given as an input to the retrieval system and the retrieval images represent muscle biological information. Although the retrieved images represent the user intent, varied anatomical structures such as upper and lower extremity are presented by the system. Again, the TBIR does not prompt CBIR action. Finally, the Query 2 and keywords "sagittal" and "left" are given as query elements to the system. These ensemble-related images to the query on the screen are based on rank order and significant features. Similarly, various keywords and query images are presented to the system and provide the system performance as listed in Table 6.8.

The overall system performance has been raised from 40% to 85.36% by adopting a modified ACO technique along with the relevance feedback mechanism. The mean average precision value of TBIR and hybrid is 78.8% and 94.81%, respectively.

6.4   SUMMARY AND CONCLUSION

The objective of the proposed framework is attained by implementing text-, visual-, and fusion-based system approaches to retrieve a heterogeneous category of medical images from the database. For this implementation, medical images represented with the different orientation, modality captured, biological, and anatomical information are considered. Images containing significant visual features are extracted by a modified ant colony optimization technique with relevance feedback mechanism. In addition, keywords, and labeled (or annotated) images are extracted by the TDF−IDF method. Both these textual and visual descriptors are fused by its rank order, weights, and precision value. The overall system performance using the fusion approach is attained at a rate of 94.81%. The limitation of this proposed framework is its computational complexity due to requiring feature extraction from both the image database and documents. If the size of the database increases, this

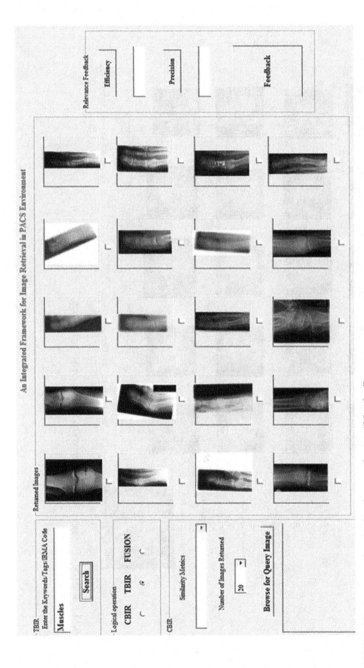

FIGURE 6.14 Images retrieved using the textual query "Muscles."

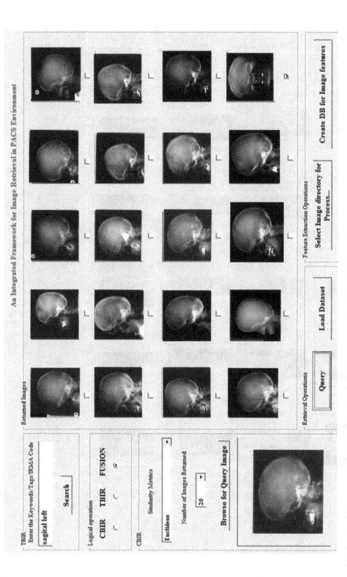

FIGURE 6.15 Retrieval of images using fusion approach for text query "sagittal left."

TABLE 6.8 Comparative Results of TBIR, CBIR and Fusion Retrieval Systems

Text	Images	$\alpha = 0$ (TBIR)		$\alpha = 1$ (CBIR)		$\alpha = 0.5$ (Fusion)	
		P (%)*	R	P (%)*	R	P (%)*	R
'arm' 'upper' 'lower'	Query 1	81	0.05	89.2	0.060	97.1	0.072
'Face' 'Sagittal' 'Left' 'right'	Query 2	80	0.05	82	0.053	91.4	0.063
'Mammography' 'Breast'	Query 3	85.5	0.07	90	0.014	96.4	0.210
'Pelvic'	Query 4	75	0.18	85.6	0.200	95.7	0.225
'Chest' 'posteroanterior' 'anteroposterior'	Query 5	72.2	0.03	86	0.034	98.6	0.039
'bone'	Query 6	81	0.10	82.8	0.104	92.1	0.113
'lower extremity' 'foot'	Query 7	75.4	0.15	84.1	0.173	94.3	0.195
'neck'	Query 8	80	0.30	87.2	0.322	92.9	0.344

P (%)*—mean average precision in percentage.

system approach would be time-consuming and may not be useful for case-based diagnosis during a crisis situation. To conquer this, the retrieval system can be implemented by more recent machine-learning algorithms, like Deep Learning.

REFERENCES

[1]   Y.D. Chun, S.Y. Seo, N.C. Kim, Image retrieval using BDIP and BVLC moments, IEEE Trans. Circuits Syst. Video Technol. 13 (9) (2003) 951–957.

[2]   A.W. Smeulders, M. Worring, S. Santini, A. Gupta, R. Jain, Content-based image retrieval at the end of the early years, IEEE Trans. Pattern Anal. Machine Intell. 22 (12) (2000) 1349–1380.

[3]   N.S. Chang, K.S. Fu, Query-by pictorial-example, IEEE Trans. Softw. Eng. 6 (6) (1980) 519–524.

[4]   H. Miller, A. Geissbuhler, S. Marchand, P. Clough, Benchmarking image retrieval applications, in: Proceedings of Tenth International Conference on Distributed Multimedia Systems (DMS'2004), 2004, pp. 334–337.

[5]   M. Flickner, H. Sawhney, H. Niblack, J. Ashley, Q. Huang, B. Dom, Query by image and video content: the QBIC system, IEEE Comput. 28 (9) (1995) 23–31.

[6]   A. Gupta, R. Jain, Visual information retrieval, Commun. ACM 40 (5) (1997).

[7]   J.R. Smith, S.F. Chang, Querying by color regions using the Visual SEEk content-based visual query system, Intell. Multimed. Inf. Retr. (1996) 23−41.

[8]   D. Daneels, D.Campenhout, W. Niblack, W. Equitz, R. Barber E. Bellon F. Fierens, Interactive outlining: an improved approach using active contours, in: Proc. SPIE Storage and Retrieval for Image and Video Databases, 1993, p. 226.

[9]   E.A. El-Kwae, H. Xu, M.R. Kabuka, Content-based retrieval in picture archiving and communication systems, J. Digit. Imaging 13 (2) (2000) 70−81.

[10] T.M. Deserno, B. Ott, 2008. IRMA (Image Retrieval in Medical Applications), <https://ganymed.imib.rwthaachen.de/irma/datasets_en.php?SELECTED = 00008#00008.dataset> (November 2008).

[11] H. Tamura, S. Mori, T. Yamawaki, Textural features corresponding to visual perception, IEEE Trans. Syst. Man Cybern. 8 (6) (1978) 460−472.

[12] C. Carson, M. Thomas, S. Belongie, J.M. Hellerstein, J. Malik, Blobworld: a system for region-based image indexing and retrieval, in: Proc. International Conference on Visual Information Systems, 1999, pp. 509−516.

[13] R.M. Haralick, B. Shanmugam, I. Dinstein, Texture features for image classification, IEEE Trans. Syst. Man Cybern. 3 (6) (1973) 610−621.

[14] O. Ludwig, D. Delgado, V. Goncalves, U. Nunes, Trainable classifier-fusion schemes: an application to pedestrian detection, in: 12th International IEEE Conference On Intelligent Transportation systems, 2009, pp. 432−437.

[15] A.L. Blum, P. Langley, Selection of relevant features and examples in machine learning, Artif. Intell. 97 (1997) 245−271.

[16] E. Frank, M. Hall, G. Holmes, R. Kirkby, B. Pfahringer, Weka − a machine learning workbench for data mining, The Data Mining and Knowledge Discovery Handbook, Springer, 2005, pp. 1305−1314.

[17] M. Gutlein, E. Frank, M. Hall A. Karwath, Large-scale attribute selection using wrappers, in: IEEE Symposium on Computational Intelligence and Data Mining, CIDM'09, 2009, pp. 332−339.

[18] L. Chen, B. Chen, Y. Chen, Image feature selection based on ant colony optimization, AI 2011: Advances in Artificial Intelligence, Lecture Notes in Computer Science, Springer, 2011, pp. 580−589.

[19] T.M. Lehmann, H. Schubert, D. Keysers, M. Kohnen , B.B. Wein, The IRMA code for unique classification of medical images, in: Proceedings of SPIE, 2003, pp. 440−445.

FURTHER READING

X. Cao, H.K. Huang, Current status and future advances of digital radiography and PACS, IEEE Eng. Med. Biol. Mag. 9 (5) (2000) 80−88.

Chapter 7

A Neuro-Fuzzy Inference Model for Diabetic Retinopathy Classification

Mohammed Imran and Sarah A. Alsuhaibani
College of Computer Science and Information Technology, Imam Abdulrahman Bin Faisal University, Dammam, Saudi Arabia

Chapter Outline

7.1 INTRODUCTION

Diabetes mellitus (DM), is a metabolic disorder in humans which recognized as one of the world's fastest growing chronic disease, and one tenth of the world's population face the risk of being diagnosed with DM. The impact of severity of DM causes diabetic retinopathy (DR), a major cause of visual impairment, which sometimes even results in permanent blindness [1]. DR accrue when glucose, the main source of energy, is not used properly by the body [2]. Diabetes was nominated as the fourth most frequently managed chronic disease in the world in 2009 and it is estimated to be the second-most frequent disease by the year of 2030 [3]. According to the International Diabetes Federation (IDF), there are 246 million persons in the world who

Intelligent Data Analysis for Biomedical Applications. DOI: https://doi.org/10.1016/B978-0-12-815553-0.00007-0

suffer from diabetes, and the number is estimated to rise to 255 million patients by 2030 [4].

DR refers to the damages and changes that happen to the retina of individuals suffering from diabetes for a long time. The disease has two main stages: the nonproliferative diabetic retinopathy stage (NPDR) and proliferative diabetic retinopathy stage (PDR). NPDR is the early stage of DR and, at this stage, the high levels of glucose in the blood make the cells of the blood vessels absorb an abnormal amount of glucose, which results in producing an additional formation of glycoproteins. The vessel's walls will then grow thicker, but weaker, which causes bleeding and leakage of fluids. This bleeding affects the fovea, the accountable area for sharp central vision. The NPDR stage is associated with the presence of three signs known as microaneurysms, hemorrhages, and exudates. The NPDR is further classified into three levels: early NPDR, moderate NPDR, and severe NPDR. During the early NPDR level, at least one microaneurysm appears in the funds retinal image, while during the moderate level multiple microaneurysms and hemorrhages appear in the retina. And in the severe NPDR, exudates appear in the retina along with the microaneurysms and hemorrhages [4,5].

PDR is a severe stage of DR; during this stage, the flow of blood is reduced in the retina and, as a result, new vessels grow to recreate blood supply. However, these vessels are fragile and grow on the surface of the retina. These new vessels may bleed, causing hemorrhage [5]. Fig. 7.1 shows a NPDR with microaneurysms, hemorrhages, and exudates appearing in the retinal fundus image.

The progression of DR can be affected by many risk factors, such as: duration of diabetes, the age of the patient, high blood pressure, poor control of diabetes, and high cholesterol level. The duration of diabetes is considered to be a major risk factor in DR development. After 10 years of diabetes infection, approximately 60% of patients suffer from DR and as diabetes' duration gets longer, the risk of developing DR increase. The age of the

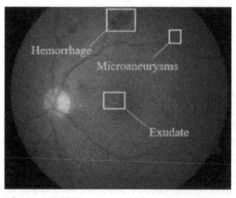

FIGURE 7.1 Retinal fundus image [6].

patient also plays a role in DR progression; older patients are at more risk to develop DR than younger patients. Uncontrolled diabetes is considered the second major risk factor in DR development after the duration of diabetes and it raises the severity of DR development [4].

The early detection of DR is essential as it plays a vital part in the treatment process and the management of the disease. It can reduce the risk of visual loss by 50%. Furthermore, making the detection process automated is important as it helps in easier detection of the DR, as the number of DR patients is increasing rapidly [7]. The main contribution of this chapter is to present a new, automated method to classify the severity of DR by using a soft-computing technique: the adaptive neuro-fuzzy inference system (ANFIS). The proposed technique will help ophthalmologists and DR patients detect DR in an effective and timely manner.

This chapter is organized as follows: In Section 7.1, provides an introduction to the topic, Section 7.2 gives a literature review of the related work. In Section 7.3, we discuss the proposed technique of ANFIS in detecting DR. Section 7.4 shows empirical studies including dataset descriptions and experimental setups. Section 7.5 presents the results and discussion, and, finally, Section 7.6 concludes with recommendations emanating from this work.

7.2   RELATED WORK

Several DR classification models were developed to detect and classify DR using different techniques. In Ref. [8], Parmar et al. proposed a model to detect DR from retinal images and classifying the images into five classes. The proposed model was based on deep convolutional neural networks. The model yielded an accuracy of 85%. In Ref. [9], Kumar et al. presented a model for retinal classification using two-field mydriatic fundus photography. Firstly, the authors located the optic disc by using multilevel wavelet decomposition and recursive region growing. Then, the blood vessels were extracted by using histogram analysis on the two median filtered images. Microaneurysms and hemorrhages were detected using three-stage intensity transformation, while exudates were detected by using multilevel histogram analysis. Finally, the extracted lesions were aggregated to classify the image as being infected by DR or not. The model yielded a sensitivity of 80% and a specificity of 50%. In Ref. [10], Xu et al. used a deep convolutional neural network method to classify diabetic retinopathy using a color fundus image. Their method yielded an accuracy rate of 94.5%. In Ref. [11], Islam et al. proposed a DR classification model based on a bag-of-words approach (BOW) with support vector machines (SVM) to classify normal and abnormal retinal. This model yielded an accuracy of 94.4%, precision of 94%, and recall of 94%. In Ref. [12], Ravishankar et al. proposed a new model to find the approximate location of the optic disk by using the intersection found from the major blood vessels detection model. They also used different morphological operations to detect DR features like exudates and microaneurysms. They evaluated the algorithm on a database of 516 images with different contrast, illumination, and disease stages. The algorithm achieved 97.1% accuracy for optic disk localization, a sensitivity and specificity of 95.7% and 94.2%, respectively, for exudate detection and 95.1% and 90.5% for microaneurysm and hemorrhage detection. In Ref. [13], Zhang et al. proposed a bright lesions detection and classification model using local contrast enhancement with fuzzy C-means and support vector machine (SVM) classifier. The proposed model achieved a sensitivity of 97% and specificity of 96% in classifying the bright lesions and bright nonlesion, sensitivity of 88% and specificity of 84% in classifying the exudates and cottonwool spots. In Ref. [14], Wang et al. proposed a classification model to classify the retinal images into two classes, normal and abnormal retinal, by classifying each pixel into two classes which are lesion or nonlesion by using the Bayesian statistical classifier. The authors achieved an accuracy of 100% in classifying all the retinal images with exudates, and 70% accuracy in classifying normal retinal images as normal. In Ref. [15], Thin et al. used the recursive region growing segmentation algorithms with the combination of the Moat Operator to classify DR. Moat operator rises the contrast between the retinal red lesions and background to makes the red lesions segmentation easier [16]. The proposed classification model yielded a sensitivity of 88.5% and a specificity of 99.7% for exudate detection, and sensitivity of 77.5% and specificity of 88.7% for hemorrhages and microaneurysms detection. In Ref. [17], Thin et al. classified the retinal changes in digital retinal images as normal, abnormal, and unknown, by first applying a local contrast enhancement technique in the preprocessing stage. Then, the retinal landmarks were eliminated from the fundus image. Finally, DR signs were recognized by using two algorithms which are recursive region growing segmentation algorithms, and a color and template matching algorithm to detect exudate and hemorrhages, respectively. In the end, all the information was collected and construed as normal, abnormal, or unknown. The proposed system yielded a sensitivity of 74.8% and specificity of 82.7%. In Ref. [18], Kahai et al. proposed the use of a decision support system (DSS) for DR classification. The authors used Bayes optimality criteria to detect microaneurysms. The proposed model yielded a sensitivity of 100% and specificity of 67%. In Ref. [19], Antal et al. introduced an ensemble-based method to detect DR. The model was built on features extracted from the output of numerous retinal image processing algorithms, like image-level including the quality and prescreening, disease signs such as microaneurysms and exudates, and retinal landmarks such as macula and the optic disc. The result concerning the existence of the disease is computed by using the ensemble approach of machine learning classifiers. The authors used the public Messidor dataset to test the model, where 90% sensitivity, 91% specificity, and 90% accuracy and 0.989 AUC was reached. In Ref. [20], Acharya et al. proposed a method to automatically identify normal, mild, moderate, severe, and prolific DR by using higher-order spectra (HOS). Firstly, the features of the row image were extracted using HOS. Then, the support vector machine (SVM) classifier was used to classify the DR level. The proposed model yielded sensitivity of 82% and specificity of 88%. In Ref. [21], Kavitha et al. proposed a model for exudates detection by using ANFIS. Features like texture and homogeneity properties and area of the exudates, were used as input to ANFIS. The classifier was tested on 200 fundus images and achieved accuracy of 99.5%.

Some researchers proposed different models to classify the severity of DR using different classification methods, but most of these methods depend on the extracted features from the retinal fundus images without taking into consideration the patient's history or the disease risk factors, such as in Ref. [22]. Choomchuay et al. proposed a model to classify the severity of DR. Firstly, the authors extracted the lesions on the retina, then the different features such as area, perimeter, and count were used with an artificial neural network (ANN) to classify the disease severity. This model achieved a classification accuracy of 96%. Also, in Ref. [23], Anidha et al. proposed a model for exudates detection and DR severity classification using ANFIS. The authors performed different prepressing techniques on the retinal images, such as green channel extraction, median filter, and histogram equalization. Then morphological operation was applied on the image. Some of the retinal features were eliminated by using the connected component technique and statistical features like exudates area, size, and color. Finally, the images were classified to the normal eye and affected eye. The model yielded accuracy of 100% in classifying retinal with pre-proliferative stage, and 97% in classifying the retinal proliferative stage.

This chapter is aimed at introducing an automated method that classifies the DR severity level into four classes using ANFIS by taking in consideration both DR signs and risk factors. To the best of our knowledge, we have not found any such work done before.

7.3 METHODOLOGY

7.3.1 Artificial Neural Network

An ANN is an electronic model designed to perform information processing in a similar way to the structure of the biological neurons in the brain. The biological neurons give the brain capability to remember, reason, and apply earlier experiences. The brain has about 100 billion neurons which are associated to each other, and the power of the human brain comes from the huge number of these neurons and the multiple connections between them [24]. The artificial neural network is a form of network that tries to mimic the brain by considering each node as artificial neurons, which gives it the ability of processing complex and nonlinear information, and is capable to work in parallel, distributed, and local processing. It also simulates the brain systems, such as the construction of architectural structures, reasoning methods, and functioning techniques [25]. The beginning of ANN was in 1943, McCulloch and Pitts have proven that there are two states of the neurons and that the state depends on threshold value. The first artificial neuron model was introduced by McCulloch and Pitts according to Ref. [26]. A lot of novel models have been developed since then. The invention of McCulloch and Pitts created a new opportunity for intelligent machines. Neural networks are applied in various areas, including the medical field, engineering, and physics, and there are two major causes for the study of ANN, namely (1) understanding the functionality of the human brain and (2) the desire to construct machines capable of solving complex problems that can't be solved by sequentially operating computers [24].

7.3.1.1   Method Explanation

In general, a biological neuron receives inputs and combines these to perform an operation on the combination, resulting in the final output. If we look at the biological neurons, there are four principal components: dendrites, cell body, axon, and synapses. The dendrites are responsible for accepting the incoming signals into the cell body. These electrical signals are then processed by the cell body and converted to the final output. The output signal is then carried out of the cell body to the other neurons through the axon, which acts as a transmission line between the neurons. The points that are located between the neurons and the dendrites are responsible for collecting the inputs from the neurons known as synapses. Fig. 7.2 illustrates the structure of the biological neurons.

The complexity of the biological neurons in the brain can be modeled, as in Fig. 7.3.

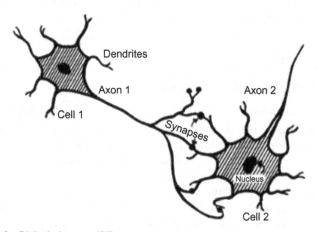

FIGURE 7.2   Biological neuron [27].

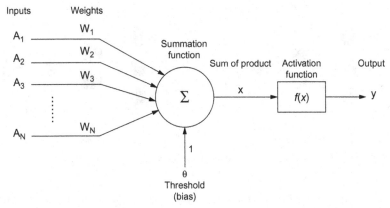

FIGURE 7.3 Artificial neuron.

In the case of ANN, the inputs enter to the body of an artificial neuron. The inputs are represented by $A(n)$; each input is multiplied with their respective weight which is a measure of the input's connection strength and represented by $W(n)$. Weighted inputs and the bias are then fed into the summing function. The value of the summing function will be sent to the activation function to generate the final output [24,28].

ANN's model can be seen in its mathematical description:

$$y(k) = F\left(\sum_{i=0}^{m} w_i(k) \cdot x_i(k) + b \right) \tag{7.1}$$

where $y(k)$ is the output, F is the activation function, $W(k)$ is the weight, $X(k)$ is the input, and b is the bias. The activation function F can be of several types, such as fixed limiter function (Eq. (7.2)), Log-Sigmoid function (Eq. (7.3)), or bipolar sigmoid function (Eq. (7.4)) [25].

$$y(k) = \begin{cases} 1, if\, x(k) \geq 0 \\ \\ 0, if\, x(k) < 0 \end{cases} \tag{7.2}$$

$$y(k) = \frac{1}{1 + e^{x(k)}} \tag{7.3}$$

$$y(k) = \frac{2}{1 + e^{x(k)}} - 1 \tag{7.4}$$

7.3.1.2 Artificial Neural Network Architecture

ANN is made of three layers namely input layer, output layer, and hidden layer/s. There must be a connection from the nodes in the input layer with the nodes in the hidden layer and from each hidden layer node with the

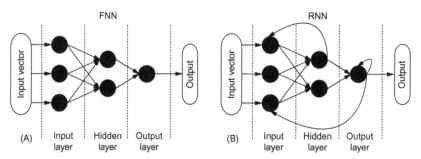

FIGURE 7.4   (A) Feedforward ANN and (B) recurrent ANN architecture [25].

nodes of the output layer. The input layer takes the data from the network. The hidden layer receives the raw information from the input layer and processes them. Then, the obtained value is sent to the output layer which will also process the information from the hidden layer and give the output. The interconnection of the nodes between the layers can be divided into two basic classes, namely the feedforward neural network and recurrent neural network. In the feedforward ANN, the information movement from inputs to outputs is only in one direction. On the other hand, in the recurrent ANN, some of the information moves in the opposite direction as well [25]. Fig. 7.4 illustrates the feedforward ANN and the recurrent ANN architecture.

7.3.2   Fuzzy Inference System

Fuzzy inference entails the processes of mapping a set of inputs to the output using fuzzy logic. Fuzzy Logic is a multivalued logic that allows an intermediate value to be defined between 0 and 1, which contrasts with the classical Boolean logic where only two values are allowed, either 0 or 1 [29]. FIS is applied in many fields, such as data classification, decision analysis, and expert systems. It is built based on three main components: membership function, fuzzy logic operators, and if−then rules [30].

7.3.2.1   Fuzzy Inference System Architecture

The architecture of FIS involves four steps: Fuzzification, Knowledge Base, Inference Engine, and Defuzzification. Fig. 7.5 illustrates the FIS architecture:

As shown in the figure, in the fuzzification step the crisp inputs are converted to linguistic variables by using the membership functions which are stored in the knowledge base. The membership function can be of different

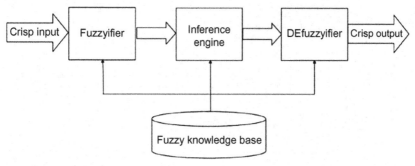

FIGURE 7.5   FIS architecture.

types, the Triangular function (Eq. (7.5)), Trapezoidal function (Eq. (7.6)), or Gaussian function (Eq. (7.7)).

$$\mu_A(x) = \begin{cases} 0, & x \le a \\ \dfrac{x-a}{m-a}, & a < x \le m \\ \dfrac{b-x}{b-m}, & m < x < b \\ 0, & x \ge b \end{cases} \tag{7.5}$$

$$\mu_A(x) = \begin{cases} 0, & (x < a) \ or \ (x > b) \\ \dfrac{x-a}{b-a}, & a \le x \le b \\ 1, & b \le x \le c \\ \dfrac{d-x}{d-c}, & c \le x \le d \end{cases} \tag{7.6}$$

$$\mu_A(x) = e^{-\frac{(x-m)^2}{2k^2}} \tag{7.7}$$

In the inference engine, the IF−THEN rules stored in the knowledge base are used to compute the fuzzy output from the fuzzy input as illustrated in Fig. 7.6.

In the defuzzification step, the fuzzy output is converted to crisp value by using the same membership function used in the fuzzification step. Different defuzzification methods can be used to obtain the crisp value, such as the Centroid of Area, Bisector of Area, Mean of Max, Smallest of Max, and Largest of Max [30].

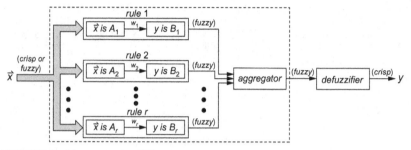

FIGURE 7.6 Inference engine [31].

7.3.2.2 Fuzzy Inference System Types

There are several types of FIS, and the most two types commonly used are the Mamdani Fuzzy model and Sugeno Fuzzy model. The difference between these types comes from the consequents, of their fuzzy rules which make the procedures of their aggregation and defuzzification different as well. A brief description of each type is discussed next.

7.3.2.2.1 Mamdani Fuzzy Model

The Mamdani method is considered the most generally used FIS technique. The method was first introduced by Professor Ebrahim Mamdani in 1975, when he applied a set of fuzzy rules supplied by experienced human operators to build the first fuzzy system to control a steam engine and boiler combination. The process of the Mamdani method is performed in six steps. First, the determination of the needed fuzzy rules. Then, (2) there is the conversion of the crisp inputs to the fuzzy inputs by using the membership function. Then, (3) the fuzzified inputs must be aggregated based on the fuzzy rules. Then, (4) the consequent of the rule is determined by joining the rule strength and (4) the output of the membership function. Next, (5) all the consequents must be combined to get the output distribution. Finally, (6) output distribution is defuzzied [32]. Fig. 7.7 illustrates the Mamdani Fuzzy Inference System.

7.3.2.2.2 Sugeno Fuzzy Model

This fuzzy method was presented in 1985 by Takagi, Sugeno, and Kang. It is very similar to the Mamdani method except that the consequent of the rule is changed where a mathematical function is used for the input variable instead if fuzzy set [30]. The general format of the rule for the Sugeno fuzzy method is:

$$\text{IF} \quad x \text{ is } A \quad \text{AND} \quad y \text{ is } B \quad \text{THEN} \quad z \text{ is } f(x, y)$$

where x, y, and z are linguistic variables; A and B are fuzzy sets; and $f(x, y)$ is the mathematical function.

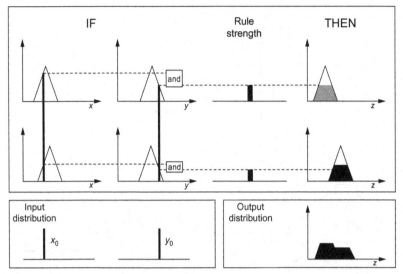

FIGURE 7.7   Mamdani fuzzy interface system [33].

7.3.3   Adaptive Neuro-Fuzzy Inference System

ANFIS is the combination of the two soft-computing methods: ANN and Fuzzy Inference System, which was first introduced by Jyh-Shing Roger Jang in 1992. It works in Sugeno fuzzy inference system and its structure is similar to the multilayer feedforward neural network structure, except that the links in ANFIS indicate the signals' flow direction and there are no associated weights with the links. Fig. 7.8A illustrates the Sugeno fuzzy model with two rules along with (B) the corresponding ANFIS architecture. To simplify the concept, two rules in the method of "If−Then" for the Sugeno model will be considered with x and y as inputs and f as output [25]. The rules are as follows:

Rule1: if x is A_1 and y is B_1, then $f_1 = p_1x + q_1y + r_1$

Rule2: if x is A_2 and y is B_2, then $f_2 = p_2x + q_2y + r_2$

where A_1, B_1, A_2, and B_2 are the membership function and p_1, q_1, r_1, p_2, q_2, and r_2 are the design parameters that are determined through the training process.

7.3.3.1   Adaptive Neuro-Fuzzy Inference System Architecture

The architecture of ANFIS consists of five layers, as illustrated in Fig. 7.8. The first and fourth layers contain an adaptive node. The other layers contain nonadoptive nodes (fixed node). Next is a brief explanation for each layer:

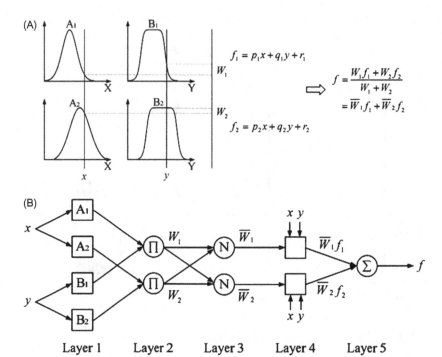

FIGURE 7.8   (A) Sugeno fuzzy model and (B) the equivalent ANFIS architecture [25].

Layer 1: Each node in this layer adjusts to a function parameter. The output of each node is a value of the membership degree [25]. The membership function of different types was discussed in Section 7.3.2.1.

$$O_{1,i} = \mu_{Ai}(x), \ i = 1, 2 \tag{7.8}$$

$$O_{1,i} = \mu_{Bi-2}(x), \ i = 3, 4 \tag{7.9}$$

where μ_{Ai} and μ_{Bi-2} are the degree of membership functions for the fuzzy sets A_i and B_i.

Layer 2: All the nodes in this layer are fixed and represent a firing strength for each rule. The output of the node is the outcome of multiplying the signals coming into the node and carried out to the next node [25].

$$O_{2,i} = w_i = \mu_{Ai}(x) \times \mu_{Bi}(y), \ i = 1, 2 \tag{7.10}$$

where w_i represents the firing strength for each rule.

Layer 3: All the nodes in this layer are fixed. The output of every node in this layer is the normalized firing strength which is the ratio between the ith rules' firing strength and the total sum of all the rules' firing strengths [25].

$$O_{3,i} = \overline{w}_i = \frac{w_i}{\sum_i w_i} \qquad (7.11)$$

Layer 4: All the nodes in this layer are adaptive nodes to an output. The node function defined as:

$$O_{4,i} = \overline{w}_i f_i = w_i(p_i x + q_i y + r_i) \qquad (7.12)$$

where $\overline{w}_i$ is the normalized firing strength from the previous layer (3rd layer) and $(p_i x + q_i y + r_i)$ are the parameters in the node [25].

Layer 5: in this layer there is only a single node which is fixed and calculates the total summation of all the arriving signals from the previous node to find the final output [24].

$$O_{5,i} = \sum_i \overline{w}_i f_i = \frac{\sum_i w_i f_i}{\sum_i w_i} \qquad (7.13)$$

7.4 EMPIRICAL STUDIES

7.4.1 Dataset Description

The dataset used in this work was built by the authors with the help of ophthalmologists from the publicly available DIARETDB0 dataset. The DIARETDB0 dataset contains 89 color fundus images which were taken in the Kuopio university hospital [6]. The number of instances in the new dataset is 89 with 11 plus the class (all numeric-valued) features with no missing values. Each instance in the dataset represent an estimated value for DR risk factors and symptoms for each retinal image in the DIARETDB0 dataset. The risk factors considered in this experiment include diabetes duration, age, sugar level, blood pressure, cholesterol level, and hemoglobin level. The DR symptoms are microaneurysms, hemorrhages, and exudates. Table 7.1 and Fig. 7.9 illustrate the features of the dataset and the class distribution respectively.

7.4.2 Statistical Analysis of the Dataset

Table 7.2 illustrates the statistical analysis of all the features in the dataset where the mean, median, standard deviation, maximum value, and minimum value are shown. Table 7.3 illustrates the correlation coefficient of each feature along with the target.

7.4.3 Experimental Setup

The experiment was performed using n Intel Core i7-6700T processor with 2.80 GHz speed. The software used for the experiment implementation is

TABLE 7.1 Features of the DIARETDB0 Dataset

Feature ID	Feature Name	Description
F1	Duration (years)	Duration of diabetes infection
F2	Age (years)	Age of the patient
F3	A1C	Hemoglobin A1C percentage
F4	Blood pressure: Systolic	blood pressure in case of systolic
F5	Blood pressure: Diastolic	blood pressure in case of diastolic
F6	Cholesterol level	—
F7	HB blood level	Hemoglobin blood level
F8	Micro level (0, 1, 2, or 3)	0 = Don't exists, 1 = low, 2 = moderate, 3 = high
F9	Hemorrhages level (0, 1, 2, or 3)	0 = Don't exists, 1 = low, 2 = moderate, 3 = high
F10	Exudates level (0, 1, 2, or 3)	0 = Don't exists, 1 = low, 2 = moderate, 3 = high
T (target)	Damage level (0, 1, 2, or 3)	0 = Not damaged, 1 = low, 2 = moderate, 3 = high

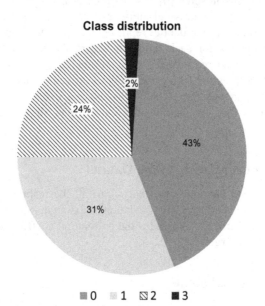

Class distribution

2%
24%
43%
31%

■ 0   ▧ 1   ▨ 2   ■ 3

FIGURE 7.9 Class distribution.

TABLE 7.2 Statistical Analysis of the Dataset

	Mean	Median	Standard Deviation	Maximum	Minimum
F1	9.98	12	8.936	25	0
F2	47.438	50	10.6	65	20
F3	7.482	7.5	2.238	13	4.5
F4	134.719	133	13.007	180	10
F5	82.775	81	7.737	105	71
F6	173.124	170	50.775	253	100
F7	14.816	15	1.287	17	11
F8	0.764	1	0.769	3	0
F9	0.562	0	0.839	3	0
F10	0.798	0	1.013	3	0
T	0.978	1	0.977	3	0

TABLE 7.3 Correlation Between Each Attribute and the Target Attribute

Attributes' Pairs	Correlation Coefficient
F1 and T	0.8459
F2 and T	0.50364
F3 and T	0.848905
F4 and T	0.166761
F5 and T	− 0.24428
F6 and T	− 0.74875
F7 and T	0.144922
F8 and T	0.870297
F9 and T	0.681375
F10 and T	0.833433

MATLAB with ANFIS classifier. Initially, the preprocessing techniques were employed on the newly created dataset. This stage includes data normalization where all the data of the dataset's features are transformed to a

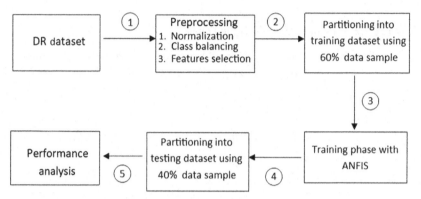

FIGURE 7.10   Flow diagram of the proposed model.

notionally common small-scale interval like 0 to +1. Then, the distribution of the classes in the dataset were balanced using the Synthetic Minority Over-sampling technique which makes the number of instances for class 0 as 36, class 1 is 26, class 2 is 20, and class 4 is 28. By using this method, the total number of the data sample in the dataset rises to 110. Then, the number of the features in the dataset were reduced using two different feature selection techniques which are the correlation coefficient and principal component analysis (PCA). Reducing the features makes the training time shorter and avoids the curse of dimensionality. Using the correlation coefficient, half of the features with the highest correlation with the target feature were selected. As illustrated in Table 7.3, the five features with the highest correlation are F8, F3, F1, F10, and F6; those features were used in this experiment. Using the PCA technique gets the most important information from the data by creating new orthogonal features that represent this information, called principal components [34]. Only the best five features were used. Finally, the dataset was spliced into two datasets, namely training and testing, using a direct partitioning method with a partitioning ratio of 60:40 for training and testing, respectively. Fig. 7.10 illustrates the flow diagram of the proposed model.

7.5   RESULT AND DISCUSSION

Two ANFIS classification models were built based on the different feature sub-sets obtained from the two feature selection methods. The classifiers were trained using a 66 data sample which represents 60% of the dataset and the two models were tested using a 44 data sample which is 40% of the dataset. The following subsections illustrate the result of the two classifiers.

7.5.1   Adaptive Neuro-Fuzzy Inference System Classifier With Correlations Coefficient Feature Selection

As illustrated in Section 7.3, half of the features with the highest correlation confection with the target feature were selected to train and test the classifier. The structure of ANFIS used in this experiment consists of five inputs and a single output. The number of epochs (iterations) used is 100 with an error tolerance of 0.0001. The first and the fifth inputs were given two membership functions, the second and the fourth were given four membership functions and the third was given three membership functions, with total number of 192 fuzzy rules. Different types of membership functions were used, and their results are shown in Table 7.4.

From the table, the highest classification accuracy was achieved by using the Triangular MF with 84.091%. The second highest accuracy level was achieved by using the generalized bell function with 81.82%, then the Gaussian MF with 77.27% accuracy. The Pi MF gave the lowest accuracy level at 75%. Fig. 7.11 illustrates the structure of ANFIS. The initial and final membership functions for the input 3 are as shown in Figs. 7.12 and 7.13. And the final decision surface is shown in Fig. 7.14.

7.5.2   Adaptive Neuro-Fuzzy Inference System Classifier With PCA Feature Selection

As illustrated in Section 7.3, the features sub-set was obtained by taking the best five features with PCA. The structure of ANFIS consists of five inputs and single output. The number of epochs (iterations) used in this experiment is 100 with error tolerance 0.0001. Each input was given two membership functions with a total number of 32 fuzzy rules. Different types of membership functions were used, and their results are shown in Table 7.5.

From the table, the highest classification accuracy was achieved by using the Triangular MF with 65.9%. The second highest accuracy level was achieved by using the Trapezoidal MF function and the fusion function with 61.3%, then the generalized bell MF with 56.8% of accuracy. The Pi MF gave the lowest accuracy level, at 54.5%. Fig. 7.15 illustrates the structure of

TABLE 7.4 Classification Result With Correlation Coefficient Features Selection

MF Type	Triangular MF	Trapezoidal MF	Gbell MF	Gaussian MF	Pi MF
Accuracy	84.09%	72.72%	81.81%	77.27%	75%

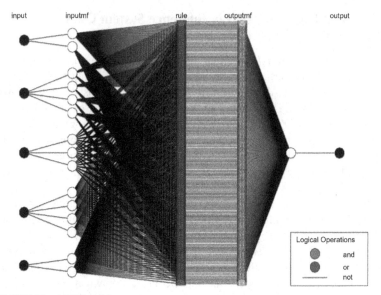

FIGURE 7.11   ANFIS structure.

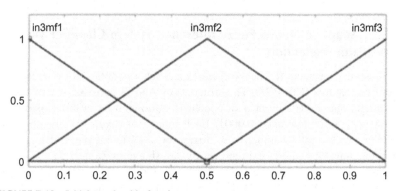

FIGURE 7.12   Initial membership function.

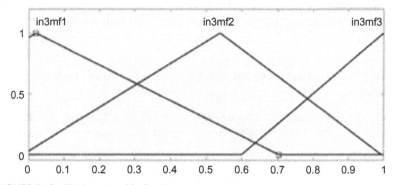

FIGURE 7.13   Final membership function.

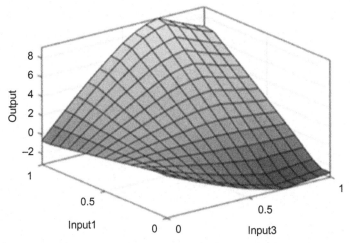

FIGURE 7.14   Final decision surface.

TABLE 7.5 Classification Result With PCA Features Selection

MF Type	Triangular MF	Trapezoidal MF	Gbell MF	Gaussian MF	Pi MF
Accuracy	65.90%	61.36%	56.81%	61.36%	54.54%

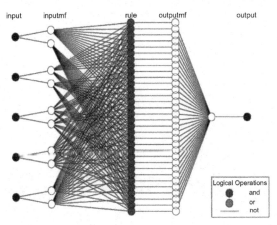

FIGURE 7.15   ANFIS structure.

ANFIS. The initial and final membership functions for the input 3 are as shown in Figs. 7.16 and 7.17. And the final decision surfaces are given in Fig. 7.18.

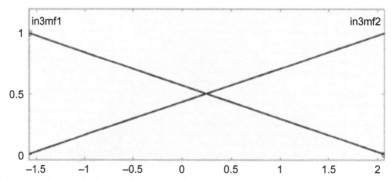

FIGURE 7.16 Initial membership function.

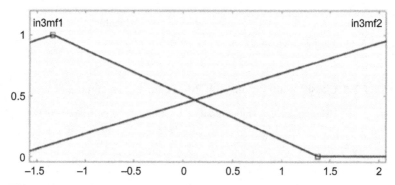

FIGURE 7.17 Final membership function.

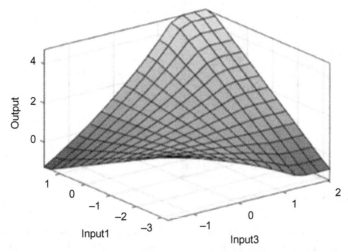

FIGURE 7.18 Final decision surface.

7.5.3   Performance Analysis

Several evaluation metrics were used to evaluate and compare the performance of the two classifiers with the highest accuracy. Fig. 7.19 below shows the pest result obtained by both sub-set features selection using PCA and the correlation coefficient. From the results, the correlation coefficient gave better classification accuracy, 84%, than PCA, with 65.90% accuracy.

Fig. 7.17 shows the RMSE for both classifiers. RMSE is a performance measure method that is used to measure the dissimilarity between the result predicted by a classifier and the target value [35]. RMSE is calculated as:

$$\text{RMSE} = \sqrt{\frac{\sum_{i=1}^{n} (y_i - \hat{y_i})^2}{n}} \tag{7.14}$$

where y_i represents the target value, $\hat{y_i}$ represents the predicted value by the classifier and n represents the number of the data sample. Fig. 7.20 provides the RMSE values for each classifier. From the result, the correlation coefficient gave less RMSE than PCA with 0.476 error, while PCA gave 0.738.

Fig. 7.18 shows the Kappa coefficient for the both classifiers. Kappa coefficient is a commonly used method in evaluating the categorical agreement between different raters or methods [35]. It is calculated as:

$$k = \frac{P_o - P_e}{1 - P_e} \tag{7.15}$$

where P_o represents the observed agreement among methods (accuracy). P_e represents the hypothetical probability of chance agreement. If $\kappa = 1$, the methods or raters totally agree with each other. If $\kappa = 0$, then the methods or raters do not agree with each other. Fig. 7.21 provides the Kappa coefficient

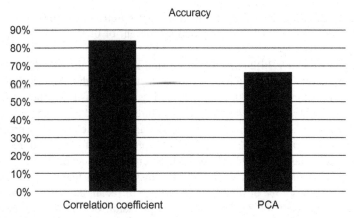

FIGURE 7.19 Classifiers accuracy.

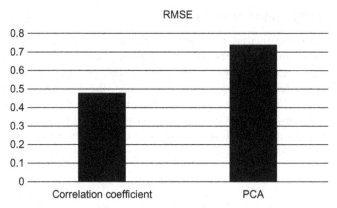

FIGURE 7.20   Classifiers RMSE.

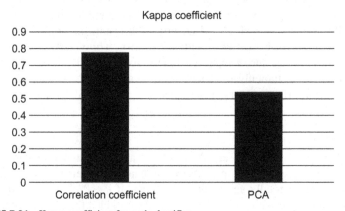

FIGURE 7.21   Kappa coefficient for each classifier.

for each classifier. From the result, the correlation coefficient gave better Kappa coefficient than PCA with total agreement of 0.791, while PCA gave 0.544.

Tables 7.6 and 7.7 illustrate the confusion matrix for the ANFIS classifier with the correlation coefficient feature selection and the ANFIS classifier with PCA feature selection respectively. Fig. 7.22 illustrates the precision and recall for the ANFIS classifier with features selected by correlation coefficient. Class 0 has 100% for both precision and recall, class 1 has precision of 50% and recall of 83.3%, class 2 has precision of 75% and recall of 37.5%, while class 3 has 92.8% for both precision and recall. Fig. 7.23 illustrates the precision and recall for ANFIS classifier with features selected by PCA. Class 0 has 75% of precision and 81.8% of recall, class 1 has precision of 53.8% and recall of 85.3%, class2 has precision of 55.5% and recall of 62.5% while class 3 has precision of 80% and recall of 61.5%.

TABLE 7.6 Confusion Matrix for Classifier 1

	0	1	2	3
0	16	0	0	0
1	0	5	1	0
2	0	4	3	1
3	0	1	0	13

TABLE 7.7 Confusion Matrix for Classifier 2

	0	1	2	3
0	9	2	0	0
1	3	7	1	1
2	0	2	5	1
3	0	2	3	8

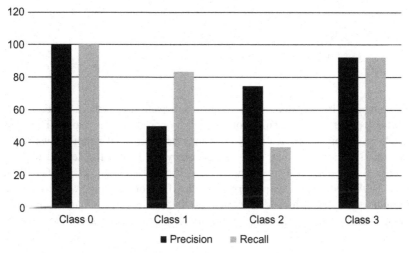

FIGURE 7.22 Precision and recall for classifier 1.

7.5.3.1   Comparison With Various Algorithms

Table 7.8 provides a comparison between the proposed work with some other recent works published in 2017.

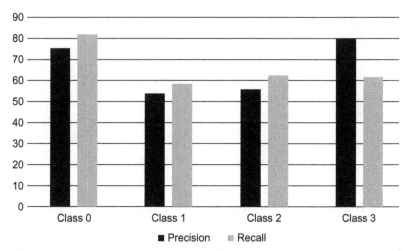

FIGURE 7.23   Precision and recall for classifier 2.

TABLE 7.8 Comparison With Various Algorithms

References	Applied Technique	Accuracy
Parmar et al. [8]	Deep convolutional neural networks	85%
Xu et al. [10]	ANN	94.5%
Islam et al. [11]	BOW with SVM	94.4%
This proposed approach	ANFIS	84.09%

As shown in the table, all the other classification models in the comparison outperform the classification model proposed in this work. The best result was achieved by Xu et al. [10], they used artificial neural network for the classification and they achieved accuracy of 94.5%. the second highest accuracy was achieved by Islam et al. [11] with accuracy of 94.4%. They used BOW with SVM to perform the classification. Finally, work done by Parmar et al. [8] achieved accuracy of 85%. It should be noted that the gained results from each work are not directly comparable since the datasets differ.

7.6   CONCLUSION

In this chapter, we explored the application of the ANFIS classifier for DR severity classification. The proposed method can offer the ophthalmologists an effective way to detect DR which will help in DR management.

The ANFIS classifier along with features selected by using the correlation coefficient shows a good performance to classify DR severity with accuracy of 84.09%. In the future, the architecture of ANFIS will be improved to achieve a higher accuracy level. Furthermore, different classification algorithms can be used to classify the same dataset.

REFERENCES

[1] N.P. Emptage, S. Kealey, F.C. Lum, S. Garratt, Preferred practice pattern: diabetic retinopathy, American academy of ophthalmology, (2016), pp. 13–19.

[2] Diabetes Steering Committee, Diabetes: a Manitoba strategy, Manitoba Health, 1998, p. 96.

[3] P. Prentasic, S. Loncaric, Z. Vatavuk, G. Bencic, M. Subasic, T. Petkovic, et al., Diabetic retinopathy image database (DRiDB): a new database for diabetic retinopathy screening programs research, in: 2013 8th International Symposium on Image and Signal Processing and Analysis (ISPA), 2013, pp. 711–716.

[4] Royal College of Ophthalmologists, The Royal College of Ophthalmologists Diabetic Retinopathy Guidelines, vol. 2012, no. July, 2013.

[5] P.H. Scanlon, Diabetic retinopathy, Medicine (United Kingdom) 43 (1) (2015) 13–19.

[6] T. Kauppi, V. Kalesnykiene, J.-K. Kamarainen, L. Lensu, I. Sorri, A. Raninen, et al., The DIARETDB1 diabetic retinopathy database and evaluation protocol, Proc. Br. Mach. Vis. Conf. 2007 (2007) 15.1–15.10.

[7] S.M. Habashy, Identification of diabetic retinopathy stages using fuzzy C-means classifier, Int. J. Comput. Appl. 77 (9) (2013) 7–13.

[8] R. Parmar, R. Lakshmanan, Detecting diabetic retinopathy from retinal images using CUDA deep neural network, Int. J. Intell. Eng. Syst. 10 (4) (2017) 284–292.

[9] P.N.S. Kumar, R.U. Deepak, A. Sathar, V. Sahasranamam, R.R. Kumar, Automated detection system for diabetic retinopathy using two field fundus photography, Proced. Comput. Sci. 93 (2016) 486–494.

[10] K. Xu, D. Feng, H. Mi, Deep convolutional neural network-based early automated detection of diabetic retinopathy using fundus image, Molecules 22 (12) (2017).

[11] M. Islam, A.V. Dinh, K.A. Wahid, Automated diabetic retinopathy detection using bag of words approach, J. Biomed. Sci. Eng. 10 (5) (2017) 86.

[12] S. Ravishankar, A. Jain, A. Mittal, Automated feature extraction for early detection of diabetic retinopathy in fundus images, IEEE Conference on Computer Vision and Pattern Recognition, IEEE, Miami, FL, USA, 2009, pp. 210–217.

[13] Z. Xiaohui, A. Chutatape, Detection and classification of bright lesions in color fundus images, 2004 International Conference on Image Processing. ICIP'04, vol. 1, IEEE, Singapore, 2004, pp. 139–142.

[14] H.W.H. Wang, W.H.W. Hsu, K.G.G.K.G. Goh, M.L.L.M.L. Lee, An effective approach to detect lesions in color retinal images, Proc. IEEE Conf. Comput. Vis. Pattern Recognition. CVPR 2000 (Cat. No. PR00662), vol. 2, 2000, pp. 1–6.

[15] C. Sinthanayothin, J.F. Boyce, T.H. Williamson, H.L. Cook, E. Mensah, S. Lal, et al., Automated detection of diabetic retinopathy on digital fundus images, Diabet. Med. 19 (2) (2002) 105–112.

[16] J. Jose, J. Kuruvilla, Detection of Red Lesions and Hard Exudates in Color Fundus Images, vol. 3, no. 10, 2014, pp. 8583–8588.

[17] A. Singalavanija, J. Supokavej, P. Bamroongsuk, C. Sinthanayothin, S. Phoojaruenchanachai, V. Kongbunkiat, Feasibility study on computer-aided screening for diabetic retinopathy, Jpn. J. Ophthalmol. 50 (4) (2006) 361–366.

[18] P. Kahai, K.R. Namuduri, H. Thompson, A decision support framework for automated screening of diabetic retinopathy, Int. J. Biomed. Imaging 2006 (2006) 1–8.

[19] B. Antal, A. Hajdu, An ensemble-based system for automatic screening of diabetic retinopathy, Knowl. Based Syst. 60 (2014) 20–27.

[20] R. Acharya U, C.K. Chua, E.Y.K. Ng, W. Yu, C. Chee, Application of higher order spectra for the identification of diabetes retinopathy stages, J. Med. Syst. 32 (6) (2008) 481–488.

[21] S. Kavitha, K. Duraiswamy, Adaptive neuro-fuzzy inference system approach for the automatic screening of diabetic retinopathy in fundus images, J. Comput. Sci. 7 (7) (2011) 1020–1026.

[22] G. Anidha, Adaptive neuro fuzzy inference system assisted diagnosis of diabetic retinopathy from fundus image, Int. J. Comput. Appl. (2013) 32–36.

[23] M.P. Paing, S. Choomchuay, M.D. Rapeeporn Yodprom, Detection of lesions and classification of diabetic retinopathy using fundus images, 2016 9th Biomedical Engineering International Conference (BMEiCON), IEEE, Laung Prabang, Laos, 2017.

[24] B.C. Bangal, Automatic Generation Control of Interconnected Power Systems Using Artificial Neural Network Techniques, Bharath University, 2010, pp. 74–100.

[25] W. Suparta, K.M. Alhasa, Modeling of Tropospheric Delays Using ANFIS, no. 2009, 2016, pp. 5–19.

[26] R. Rojas, Neural networks: a systematic introduction, Springer Science & Business Media, 2013, pp. 5–9.

[27] Thinking About the Brain: The Institute for Creation Research. Available from: http://www.icr.org/article/thinking-about-brain/ (accessed 18.02.16).

[28] A. Krenker, J. Bešter, A. Kos, Introduction to the artificial neural networks, Eur. J. Gastroenterol. Hepatol. 19 (12) (2011) 1046–1054.

[29] Franck Dernoncourt, Introduction to fuzzy logic control, Essentials Fuzzy Model. Control (January) (2013) 109–153.

[30] F. Cavallaro, A Takagi-Sugeno fuzzy inference system for developing a sustainability index of biomass, Sustainability 7 (9) (2015) 12359–12371.

[31] H. Jalalifar, S. Mojedifar, A.A. Sahebi, Prediction of rock mass rating using fuzzy logic and multi-variable RMR regression model, Int. J. Min. Sci. Technol. 24 (2) (2014) 237–244.

[32] I. Iancu, A Mamdani type fuzzy logic controller, Fuzzy Log. Control. Concepts Theor. Appl. (2012).

[33] M.M.S. Mekhail, Impact of allocation the resources on decision-making in course of planning processes through design predictive model using fuzzy logic approach, J. Build. Constr. Plan. Res. 5 (4) (2017) 146–158.

[34] H. Abdi, L.J. Williams, Principal component analysis, Wiley Interdiscip. Rev. Comput. Stat. 2 (4) (2010) 433–459.

[35] S. Ghosh, S. Biswas, D. Chanda (Sarkar), P.P. Sarkar, Breast cancer detection using a neuro-fuzzy based classification method, Indian J. Sci. Technol. 9 (14) (2016).

Chapter 8

Computational Automated System for Red Blood Cell Detection and Segmentation

Muhammad Mahadi Abdul Jamil[1], Laghouiter Oussama[1],
Wan Mahani Hafizah[1], Mohd Helmy Abd Wahab[2] and
Mohd Farid Johan[3]

[1]*Department of Electronic Engineering, Faculty of Electrical and Electronics Engineering, Universiti Tun Hussein Onn Malaysia, Pt. Raja, Malaysia,* [2]*Department of Computer Engineering, Faculty of Electrical and Electronics Engineering, Universiti Tun Hussein Onn Malaysia, Pt. Raja, Malaysia,* [3]*Department of Hematology, School of Medical Sciences, Universiti Sains Malaysia, Kubang Kerian, Malaysia*

Chapter Outline

8.1 INTRODUCTION

Hematology is a branch of internal medicine science concerning the diagnosis, treatment, and prevention of diseases related to blood [1]. Blood is defined as a fluid connective tissue made of 55% plasma and 45% blood cells. Blood cells are comprised of red blood cells (RBCs), or erythrocytes, which are derived from erythroblasts in the bone marrow and are packed

Intelligent Data Analysis for Biomedical Applications. DOI: https://doi.org/10.1016/B978-0-12-815553-0.00008-2

with hemoglobin, white blood cells (WBCs, leukocytes), and platelets (thrombocytes) [2]. The different types of blood cells have specific functions in human body: The main role of RBC is to carry oxygen into deep tissues; the WBCs protect the body against foreign microbes and toxins; and platelets clot the blood in the body after injuries to protect the body from blood loss [2].

Human blood is highly sensitive and prone to various diseases and infections which can affect the function and shape of the blood cells and can endanger human health. One of the blood diseases is anemia. The World Health Organization (WHO) in 2015 declared that anemia is one of the most dangerous illnesses worldwide. The condition of anemia can regularly be gained based on medical records and examination of blood. For example, if there is a record of blood donor rejection, the record should be thoroughly checked as it could be the sign of the patient's anemic condition that has been examined. The historical record of the entire family must also be known especially their past medical treatment, medicine or drugs that have been used, hobbies, work, and any exposure to hazardous conditions to assist in the investigation of possible bleeding disorders, anemia, splenectomy, jaundice, abnormal hemoglobin, and cholelithiasis.

8.2 COMPLETE BLOOD COUNT

As a preventive step to detect anemia in an early stage, a blood test is required periodically and regularly examined by a clinic [3]. The clinical examination determines the full blood count (FBC) or complete blood count (CBC). CBC involves three blood cell type tests (i.e., RBC, WBC, and platelets) and some measurement values like mean cell hemoglobin (MCH), red cell distance width (RDW), and mean cell volume (MCV). During periodic health examinations, the various blood counts show a reading with the reference range. It is a sensitive barometer of numerous illnesses, and the measurement of the examination is an essential part of regular periodic health examinations [4]. Any abnormal reading could be the result of infections, diseases, or other illnesses.

The RBC count is one part of a CBC examination and is performed to diagnose anemia, polycythemia, and other disorders related to the RBCs, such as thalassemia and leukemia [5]. Thus, RBC count is a crucial clue in detecting many illnesses. Conventionally, the number of RBCs in a blood image were counted manually using a microscope. Manual counting is tedious, time-consuming, and prone to human error. The accuracy of manual counting is subject to the mood and visual inspection capability of the analyzer to count the cells using a hemocytometer. Other conditions that can hamper the accuracy of cell counting are poor illumination, noise, and weak edge boundaries of the image. To overcome these obstacles, automated RBC counting devices were invented in 1950 as an alternative method for use in clinical laboratories [6]. These devices are capable of producing the CBC values, including RBC count, accurately. However, in the case of agglutination, doublet erythrocytes are considered as one RBC, while larger clumps are not counted as RBC [7]. As a result, this leads to the decrement of RBC count and a false elevated MCV, MCH, and mean corpuscular hemoglobin concentration (MCHC).

Automatic RBC segmentation is an unconventional and accurate method that is less time-consuming in detecting and measuring RBC disorders to classify the normal and abnormal cells, since each RBC disorder shows signs of certain illnesses depending on the RBC morphology. This chapter proposes an automatic system for the detection and segmentation of RBCs with a novel approach to overcome the issue of highly overlapped cells.

8.3   RED BLOOD CELL SEGMENTATION

There has been much research concerned with the detection and segmentation of RBCs with the aim of improving the accuracy of RBC counts from a blood smear image. Various methods were proposed to eliminate the issue of separating highly overlapped cells. The main strategy for segmentation is hierarchical partitioning which operates either in inductive or deductive ways. By employing the inductive method, the stained objects can be identified, followed by classification through segmentation of the RBC regions to determine whether they are inside the RBC regions or not. In the deductive approach, the segmentation process starts at the higher-level object plane. It then proceeds to segmentation within the substructure components. In other words, this approach segments background (plasma) from the foreground (cellular objects). Advanced segmentation steps separate the RBCs from other blood cells. Then, the stained components are segmented within the RBCs [8]. There are also issues in segmentations; over-segmentation and under segmentation are serious challenges allied with the segmentation of the cell regions [9]. Under segmentation is the process in which the segmented area contains a further group of objects to be segmented. Undefined cell boundaries due to cell overlapping or firmly placed cell boundaries can result during segmentation. Over-segmentation further segments the extracted objects in the background to subcomponents. It is usually caused by a lack of understanding of the cell parameters used for segmentation. For instance, parameters such as circular shape and size of the RBCs are commonly used for RBC segmentation. However, an enlarged circular artefact or noise can affect the segmentation accuracy and lead to over-segmentation.

8.4   RELATED WORK

An inductive segmentation technique based on granulometry and automatic thresholding has been explained by Ref. [10]. In this study, granulometry was employed. Granulometry is computed from a series of morphological openings with predefined structuring elements of increasing the size to estimate the size of the cells and uses the information to determine the connective sets where the nuclei of parasites are detected using regional extrema. However, the method does not provide a detailed evaluation of segmentation. The studies were undertaken by following the deductive approach for segmentation in which a modified granulometry technique, called area granulometry, was proposed to estimate the area of the cells to differentiate the regions [11,12]. Area granulometry considers the volume of pixels removed during consecutive morphological opening operations of increasing size. The area information obtained is used to segment the image into foreground and background regions containing cellular objects and plasma, respectively. In excluding the WBC from the foreground, the method is prone to over-segmentation as it considers all stained components. The accuracy rate of granulometry is highly dependent on cell overlapping. For images containing highly overlapped cells, the method fails to accurately measure the size of the components.

A modified segmentation technique, called minimum area watershed transformation technique for automated diagnosis, was described by Ref. [13]. In this method, peaks obtained from an intensity histogram of the images provides a double threshold value which was then used for the segmentation process. For a less complex image, this process can be minimized by taking the image histogram of the original grayscale image which then yields two peaks and, thereby, the threshold for binarization. The threshold value is usually obtained by calculating the center point between the two peaks. This method utilizes the area information obtained from area granulometry such that, as the segmentation proceeds, a region will not be recognized until it has reached a certain area.

The minimum area of the watershed segmentation procedure is used as the primary segmentation operator followed by extraction of markers using Radon transform to detect the cell centers [14]. Even though the method addresses over- and under-segmentation issues to a great extent, it is not applicable to highly concentrated regions of the blood film as the cell regions cannot be clearly defined and Radon transform fails to pick up the cell centers. The granulometry technique is also not very productive for an image with heavily clumped cells.

Similar studies were conducted in which the cell boundaries were detected using Hough transform and extracted using fuzzy curve tracing [15]. A blood cell segmentation approach using evolutionary methods described the differential and artificial bee colony evolutionary methods used to perform segmentation using histogram information and a minimum distance estimator [16]. Similar ideas can be found in published research on blood cell segmentation for different applications [17]. However, all these studies lack detailed evaluations of the performance of the segmentation strategy, especially under nonuniform lighting conditions. Techniques that use inductive strategies for RBCs segmentation should elaborate the preprocessing techniques that deal with platelets and artefacts, rather than assuming all stained components as RBCs and WBC. Similarly, in the deductive approach, the differentiation of WBC from RBCs has to be addressed. Cell overlapping remains a serious issue for all these techniques.

8.5 METHODOLOGY

The methodology involved from the extraction of the blood smear image to the detection and counting of RBCs involves several stages. Fig. 8.1 illustrates the overall research framework with the steps involved from image acquisition to the separation of the overlapped cells for the proposed RBC counting. The various steps involved in the process for the preparation of the image are discussed next.

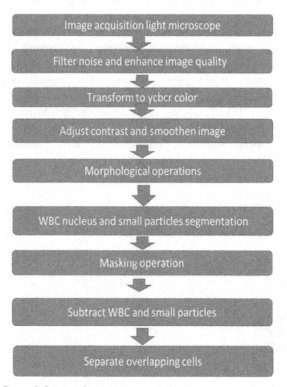

FIGURE 8.1 Research framework.

Preprocessing

Blood smear images are acquired from the blood sample using a microscope. The images are then processed and digitalized for preparation, which involves filtering out noise and enhancing the images. Conversion to a YCbCr image by adjusting the contrast and smoothing is also included to improve the image quality and increase the visibility of the cells. The morphological operation is applied before segmentation. Masks are then used for the images and cells are separated for RBC counting. A blood smear is the preparation of a blood specimen on a slide to be observed under a microscope. The process of displaying the image involves digitization of the image from the optical image. However, people cannot interpret the digitized form of the image, therefore, an interpolation process is needed to provide a continuous image from the digital image approximation to represent the optical image in the display image. In practice, the process of digitization and interpolation causes distortions and noise to the image (Fig. 8.2).

The next step for image acquisition is image digitization. Digitization is the process for displaying the RBC image which includes the digitization of the optical image with 40 times (40 ×) objective which is equal to approximately 400 times magnification, or an optical image with 100 times objective (100 ×) which requires microscopic oil on the RBC sample. Different storage methods of the raw files into computerized image formats are available in commonly used image file formats, such as graphics interchange format (GIF), joint photographic experts group (JPG), portable network graphic (PNG), bitmap (BMP), and tagged image file format (TIFF or TIF). This chapter presents the graphical user interface (GUI) of RBC segmentation which involves image conversion, image enhancement, morphological operation, and RBC segmentation and counting.

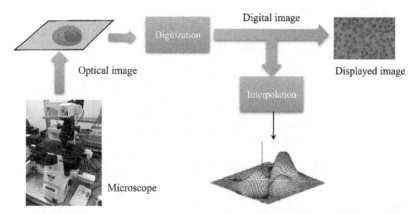

FIGURE 8.2   The flow process of image acquisition from a microscope to displayed image.

8.6   PROPOSED METHOD FOR RED BLOOD CELL DETECTION AND SEGMENTATION

The controlled watershed method with morphological operation is robust and efficient for early segmentation algorithms as reported by Refs. [13,18]. It is fundamentally different from conventional segmentation tools as it uses edge detectors to connect to pixel component extractors. The solution to rectify the under- and over-segmentation problems is by utilizing watershed segmentation to divide images into unique regions based on their regional minima. For overlap blood cell images, watershed segmentation is very effective with the use of a marker [19].

Watershed segmentation increases the architectural complexity and computational cost of the segmentation algorithm. It also successfully overcomes the problems of high overlap RBC. Fig. 8.3 shows the pseudocode of the developed marker-controlled watershed method. The neighborhood of each single pixel is defined by the Euclidian distance measurement based on the diameter of the structuring elements so that Souter and sinner return the sum of intensities of all pixels to the neighborhood of the outer and inner circular areas of the structuring elements, respectively. Two counters are employed to count the pixels charted with corresponding structuring elements during each pass. The boxed area specifies the computation of the intensity.

The pseudocode of linked components based on the marker-controlled watershed algorithm is shown in Fig. 8.3, where p represents a pixel, I is the input preprocessed image (processed by filter and morphological operation), and L is the segmented label image.

$I(p)$ defines the gray level value of p, ne is the neighbor pixel of p, and $I(ne)$ represents gray level value of the next respective pixel. The array $d[p]$ is used to store the distance from the lowest pixels or plateau. However, the array $l[p]$ is used to stock the labels. DMAX and LMAX indicate the maximum distance and maximum value for the label in the structure, respectively. DMAX determines the distance between the first pixels of the first row to the last pixel of the last row. Image scanning can be continued for step two if the $v[p]$ array structure does not change. The image scanning can continue if $l[p]$ array structure does not change.

8.7   DEVELOPED GRAPHICAL USER INTERFACE

GUI was designed to allow users to interact with the application through graphical icons and visual indicators, such as secondary notation, instead of text-based user interfaces. The GUI was developed to analyze and evaluate the concept of automated RBC segmentation. The user has the option to visually analyze the blood image, process it, and show the total RBC numbers. The GUI was created using MATLAB software package. The created

Algorithm 1 The pseudo-code of the algorithm
1: Input : f, Output : 1
2: v[p] ← 0, 1[p] ← 0, New_label ← 0, Scan_Step2 ← 1, Scan_Step3 ← 1 //
Initialization
3: Scan from top left to bottom right : STEP1(p)
4: while Scan_Step2 = 1 do
5:     Scan image from top left to bottom right : STEP2(p)
6:     if v[p] is not changed then
7:         Scan_Step2 ← 0
8:     else
9:         Scan image from bottom right to top left : STEP2(P)
10:         if v[p] is not changed then
11:             Scan_Step2 ← 0
12:         end if
13:     end if
14: end while
15: while Scan_Step3 = 1 do
16:     Scan image from top left to bottom right : STEP3(p)
17:     if 1[p] is not changed then
18:         Scan_Step3 ← 0
19:     else
20:         Scan image from bottom right to top left : STEP3(p)
21:         if 1[p] is not changed then
22:             Scan_Step3 ← 0
23:         end if
24:     end if
25: end while
26: function STEP1(p)
27:     if v[p] ≠ 1 then
28:         for each n of p // n is neighbor pixel of p
29:         if f[n] < f(p) then v[p] ← 1
30:         end if
31:     end if
32: end function

FIGURE 8.3   The pseudocode of the algorithm.

application allows any user, even those with no previous knowledge of MATLAB software or its specific tools, or the execution of complex blood image treatment tasks. It also offers interactivity and is user friendly. Using the powerful numerical computation that MATLAB provides for image processing, the application has been created for one of the CBC parameters, that is, RBC counting.

8.7.1   Application Window

Fig. 8.4 shows the graphic design software (GDS) application window with the different parts numbered for easy explanation. The left column of the GUI displays various intermediate steps and the bottom section shows the image conversion, image enhancement, and morphological operation. The top two windows of the GUI represent the image to be processed (top-left window) and the processed image with separated, marked cells (top-right window).

The descriptions of the buttons numbered from 1 to 10 in Fig. 8.4 are:

1. Application for RBC segmentation: application title.
2. Input: shows the loaded image required for processing or treatment.
3. Output: shows the result of the input image treatment with the options of image conversion, image enhancement, morphological operation, or RBC segmentation.
4. Load image: by pressing the push button to load an image, the MATLAB function directs the user to the blood image dataset path by creating a new window where you can choose the target image for processing.
5. Overlap percentage: shows the percentage of overlap pixels.
6. Gray image: by clicking the icon, the user will get a gray image which is the converted RGB blood image.
7. Image enhancement: this function uses histogram equalization to enhance the blood images.
8. Morphological operation: By pressing the push button of morphological operation call the interacted with the main GDS app.

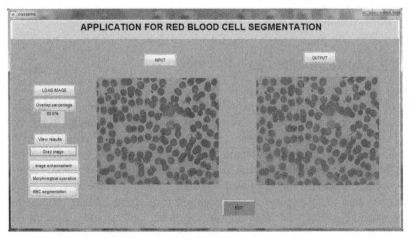

FIGURE 8.4   Graphic Design Software User Interface.

9. RBC segmentation: this function separates overlapped red blood images and shows the number before and after segmentation for each image.
10. Exit: clicking this icon will quit the application and gives the user the option to save the output image.

By clicking "yes" a new window directs the user to existing files to save the output image; by clicking "no" the user leaves the application.

Furthermore, "imsave," creates a save image tool in a separate window that is associated with the current image. The save image tool displays an interactive file chooser dialog box in which a path and filename with different extensions can be specified. When clicking save, the save image tool writes the target image to a file using the image file format selected; "imsave," uses "imwrite" to save the image, using default options.

The GDS uses the "imread" function for reading an image from the graphics file. The "imread" function is compatible with most formats, listed here in alphabetical order by format name:

- PNG: portable network graphics
- PPM: portable pixmap
- RAS: sun raster
- TIFF: tagged image file format
- JPEG: joint photographic experts group
- JPEG 2000: joint photographic experts group 2000
- PBM: portable bitmap
- PCX: windows paintbrush
- BMP: windows bitmap
- CUR: cursor file
- GIF: graphics interchange format
- HDF4: hierarchical data format
- ICO: icon file
- PGM: portable graymap

8.7.2   Application Interaction

The main feature of an app is the inherent interaction with the user; therefore, the code development is related to the software programming methods. The app is an event-driven program and once is running, it remains quiescent until the user produces an event by interacting with the app by clicking an icon or introducing text, etc. The app responds to any action by running the specified callback function that executes the pertinent code in response to the event that triggered it. So the callback is conceived as a short function that must obtain the data required to perform its task, update the app when necessary, and store its results for other callbacks to access them.

The underlying app is essentially a collection of small functions working together to accomplish a larger task. It is implied then that when writing in an event-driven program, all the actions performed over the app through the controls added must be linked to its corresponding callback, as well as the data transferring between callbacks. The apps in MATLAB can be written either programmatically using MATLAB functions or be created using GUIDE, the MATLAB Graphical User Interface Design Environment. Most users find it easier to use GUIDE to graphically layout the GUI and generate an event-driven layout for the app. Some, however, prefer the extra control they get from authoring apps from scratch. The GDS was developed using GUIDE, but for the exceptional necessities of some of the functions, it was necessary to write the code programmatically.

There are several functions that work as separate apps and share information with the GDS main app. The GDS connects with these other apps in order to set options for configuration and previsualization effects and then the resultant image or parameters are generated. Fig. 8.5 shows a block diagram of these different apps.

The RBC cell application has a user-friendly graphical interface to facilitate easy deployment of medical information for users who do not necessarily have any engineering or computer coding background. This is also important for improving and enhancing the medical diagnosis service by reducing the time and hiding complexities in the computations. The use of an appropriate graphical interface also reduces human errors and, thus, improves diagnosis accuracy. The next section provides some examples of the RBC counting application steps.

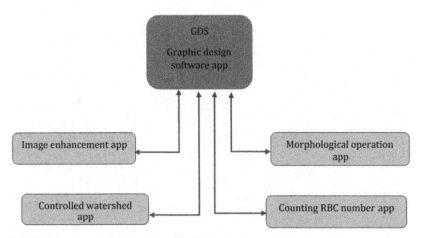

FIGURE 8.5 Graphic design software block diagram.

8.7.3 Gray Image Conversion Technique

Fig. 8.6 shows the result of the first step of image conversion. The input image is an RGB image as shown on the left side of the window while the converted grayscale image is illustrated on the right side of the window soon after the input image.

8.7.4 Image Enhancement Technique

Fig. 8.7 shows the interface during the image enhancement stage of the developed application for RBC segmentation.

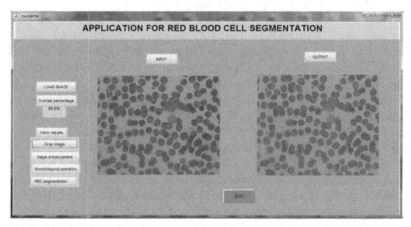

FIGURE 8.6 GUI demonstration of automated RBC segmentation setup; window displaying image conversion.

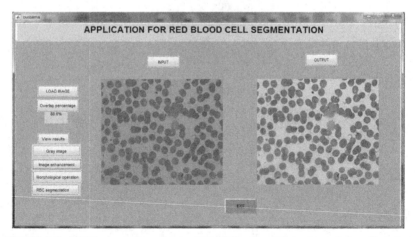

FIGURE 8.7 GUI demonstration of automated RBC segmentation setup; window displaying image enhancement step.

8.7.5 Morphological Operation Technique

Fig. 8.8 illustrates the intermediate results obtained from the morphological processing of the image. The corresponding processed image is also provided in the left column.

8.7.6 Red Blood Cells Segmentation and Counting Technique

Figs. 8.6–8.9 illustrate the main application steps for the RBC segmentation and counting. The techniques used were highly sensitive to the RBC area

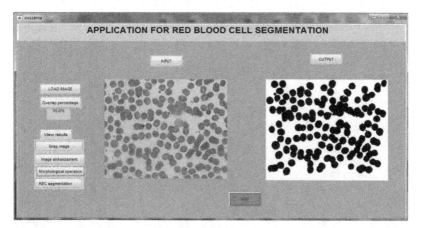

FIGURE 8.8 GUI demonstration of automated RBC segmentation setup; window displaying morphological operation.

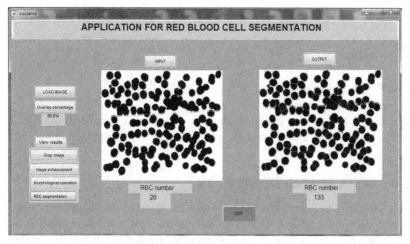

FIGURE 8.9 GUI demonstration of automated RBC segmentation setup; window displaying image segmentation step with red blood cell number.

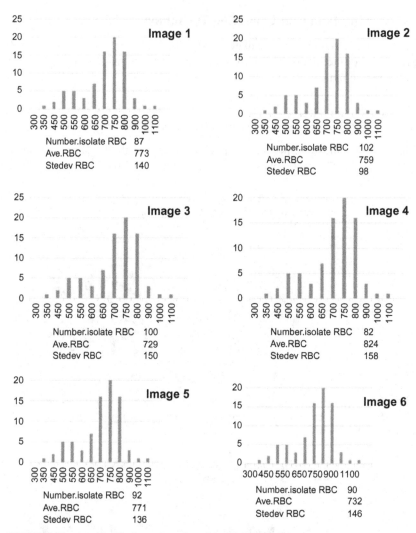

FIGURE 8.10 Area distribution and number of isolated RBC.

which can vary under different imaging conditions. The GDS app has to interact with other applications for it to be a fully functional tool.

8.8 RESULTS AND DISCUSSION

Fig. 8.9 shows segmentation results with RBC number. The controlled watershed algorithm is able to handle 15 overlapped RBCs and performed better than the geometrical feature method reported by Ref. [20], which can only deal with a maximum of five overlapped RBCs. The proposed method is less complex as it involves preprocessing techniques and the morphological erosion and dilation to remove the platelets and artefacts. Unlike conventional segmentation procedures, the proposed method provides the location information of RBC and an estimation of the cell population. The method is independent of the average level of illumination and its variations. Additionally, it performs well with images of varying illumination.

The average area of separated RBC and its robust technique are used to validate and confirm the obtained results of RBC clustering by considering the overlapping pixels between two minimum RBCs. For each blood image, there is a sufficient number of isolated RBC [21]. The control of the area of the RBC is validated by counting the separated RBCs. A normal distribution of approximately 767 ± 142 pixels of RBC is generated which implies 15.7 ± 1.3 pixels of RBC radius and the areas are within the range of $57-108$ separated RBC, which is actually sufficient to calculate the average RBC area [21]. Fig. 8.10 shows the area distribution and the number of isolated RBCs for six blood images taken randomly.

In this chapter, several images of blood cells were used to evaluate the performance of the algorithm and found sensitivity and precision values of 96.4% and 95.5%, respectively. The proposed algorithm showed an average accuracy of 95% (95% of the RBCs were detected). The algorithm successfully segmented 2419 overlapped RBCs from 2546 blood cells. Thus, the error rate of this RBC counting algorithm is only about 5%. In addition, the speed of the algorithm's results is important. To obtain these results they were written in MATLAB and run on a system with: Intel Core i7 Processor, 4-GB main memory, Microsoft Windows 7 64-bit operating system, and MATLAB™ (R2015a). The time taken by the processor to complete the diagnosis was approximately 16 s.

8.9 CONCLUSION

There are many challenging problems in automatic cytological processing of images of blood cells. Problems include the large variation of blood cells, occlusions, low quality of images, and difficulties in getting enough real data, which have all been addressed in this study. In this work, the efficient step-by-step segmentation algorithm was discussed to describe the automatic detection and segmentation of microscopic blood imagery. Experimental results indicated that the developed system offers good segmentation and recognition accuracy using random blood samples. The GUI was created using MATLAB software package. The created application is friendly user, even those with no previous knowledge of MATLAB software or its specific tools or experience with the execution of complex blood image treatment tasks, to apply the program. It also offers interactivity and user-friendliness.

Using the powerful numerical computation that MATLAB provides for image processing, the application has been created for one of the CBC parameters, that is, RBC counting. The code provides good robustness with less time during the execution of the application.

REFERENCES

[1] C. Neunert, et al., The American Society of Hematology 2011 evidence-based practice guideline for immune thrombocytopenia, Blood 117 (16) (2011) 4190−4207.
[2] M. Miswan, et al., An overview: segmentation method for blood cell disorders, in: 5th Kuala Lumpur International Conference on Biomedical Engineering 2011, Springer, 2011.
[3] M. Madjid, O. Fatemi, Components of the complete blood count as risk predictors for coronary heart disease: in-depth review and update, Texas Heart Inst. J. 40 (1) (2013) 17.
[4] C. Briggs, et al., ICSH guidelines for the evaluation of blood cell analysers including those used for differential leucocyte and reticulocyte counting, Int. J. Lab. Hematol. 36 (6) (2014) 613−627.
[5] L. Van Hove, T. Schisano, L. Brace, Anemia diagnosis, classification, and monitoring using cell-dyn technology reviewed for the new millennium, Lab. Hematol. 6 (2000) 93−108.
[6] O. Oliver, G. Lloyd, M. Isabelle, L.M. Almond, C. Kendall, K. Baxter, et al., Automated cytological detection of Barrett's neoplasia with infrared spectroscopy, J. gastroenterol. 53 (2) (2018) 227−235.
[7] C.H. Cho, et al., Measurement of RBC agglutination with microscopic cell image analysis in a microchannel chip, Clin. Hemorheol. Microcirc. 56 (1) (2014) 67−74.
[8] F.B. Tek, A.G. Dempster, I. Kale, Computer vision for microscopy diagnosis of malaria, Malaria J. 8 (1) (2009) 153.
[9] M.A. Gonzalez, G.J. Meschino, V.L. Ballarin, Solving the over segmentation problem in applications of Watershed transform, J. Biomed. Graph. Comput. 3 (3) (2013) 29.
[10] C. Di Ruberto, et al., Automatic thresholding of infected blood images using granulometry and regional extrema, in: Proceedings of 15th International Conference on Pattern Recognition, 2000, IEEE, 2000.
[11] K.M. Rao, A. Dempster, Area-granulometry: an improved estimator of size distribution of image objects, Electron. Lett. 37 (15) (2001) 950−951.
[12] K.M. Rao, A. Dempster, Modification on distance transform to avoid over-segmentation and under-segmentation, in: Video/Image Processing and Multimedia Communications 4th EURASIP-IEEE Region 8 International Symposium on VIPromCom, IEEE, 2002.
[13] F. Tek, A. Dempster, I. Kale, Blood cell segmentation using minimum area watershed and circle radon transformations, Mathematical Morphology: 40 Years On, Proceedings of the 7th International Symposium on Mathematical Morphology, April 18−20, 2005, pp. 441−454.
[14] A. Myna, M. Venkateshmurthy, C. Patil, Detection of region duplication forgery in digital images using wavelets and log-polar mapping, in: International Conference on Computational Intelligence and Multimedia Applications, 2007, IEEE, 2007.
[15] P.P. Guan, H. Yan, Blood cell image segmentation based on the Hough transform and fuzzy curve tracing, in: 2011 International Conference on Machine Learning and Cybernetics (ICMLC), IEEE, 2011.

[16] V. Osuna, E. Cuevas, H. Sossa, Segmentation of blood cell images using evolutionary methods, EVOLVE-A Bridge Between Probability, Set Oriented Numerics, and Evolutionary Computation II., Springer, 2013, pp. 299–311.

[17] L.B. Dorini, R. Minetto, N.J. Leite, Semiautomatic white blood cell segmentation based on multiscale analysis, IEEE J. Biomed. Health Inform. 17 (1) (2013) 250–256.

[18] S. Kareem, R.C. Morling, I. Kale, A novel method to count the red blood cells in thin blood films, in: 2011 IEEE International Symposium on Circuits and Systems (ISCAS), IEEE, 2011.

[19] J.M. Sharif, et al., Red blood cell segmentation using masking and watershed algorithm: a preliminary study, in: 2012 International Conference on Biomedical Engineering (ICoBE), IEEE, 2012.

[20] I. Ahmad, S. Abdullah, R.Z.A.R. Sabudin, Geometrical vs spatial features analysis of overlap red blood cell algorithm, in: International Conference on Advances in Electrical, Electronic and Systems Engineering (ICAEES), IEEE, 2016.

[21] S. Moon, et al., An image analysis algorithm for malaria parasite stage classification and viability quantification, PLoS One 8 (4) (2013) e61812.

Chapter 9

Evolutionary Algorithm With Memetic Search Capability for Optic Disc Localization in Retinal Fundus Images

B. Vinoth Kumar[1], G.R. Karpagam[2] and Yanjun Zhao[3]

[1]Department of IT, PSG College of Technology, Coimbatore, India, [2]Department of CSE, PSG College of Technology, Coimbatore, India, [3]Department of CS, Troy University, Troy, AL, United States

Chapter Outline

9.1 INTRODUCTION

Studies from the World Diabetes Foundation (WDF) reveals that patients with diabetes are at higher risk of an eye disease called diabetic retinopathy (DR) which causes damage to blood vessels in the retina leading to vision loss. DR emerges as small changes in the retinal capillaries. Although the most effective treatment is early detection through regular screenings, with advances in technology, automatic screening of these images would help doctors easily detect a patient's condition accurately. Optic disc localization is an important aspect of computer-aided diagnosis in retinal screening. Knowledge of the position of the OD may help to determine whether images are from the left or right eye and to determine the clinical relevance of lesions. The disc itself has an appearance which is different from the other structures on the retina. Parts of the disc can potentially be detected as a lesion and such false responses can be avoided if the location of the disc is known. The optic disk localization is broadly categorized to eight groups: (1) template matching

Intelligent Data Analysis for Biomedical Applications. DOI: https://doi.org/10.1016/B978-0-12-815553-0.00009-4
191

[1,2]; (2) intensity based [3,4]; (3) Hough transforms [2,5]; (4) fuzzy convergences [6]; (5) binary vasculature [7]; (6) pyramidal decomposition [8]; (7) Principal Component Analysis (PCA) [9]; (8) morphological approaches [10]; and (9) metaheuristics techniques [11,12].

This chapter starts with describing a set of terms and terminologies. The *Retina* is the light-sensitive tissue lining at the back of the eye, which converts light into electrical impulses that are sent to the brain through the optic nerve. *Diabetic retinopathy* is an eye disease that usually affects people who have had diabetes (diagnosed or undiagnosed) for a significant number of years. At first, the lesions on the retina especially, blood vessels, exudates, and microaneurysms are used along with the features such as area, perimeter, and count from these lesions for automatic detection of diabetic retinopathy. *Microaneurysm*: A tiny swelling on the side of a blood vessel which can rupture and leak blood. *Nonproliferative diabetic retinopathy*, is the earliest stage of diabetic eye disease that progresses from mild to moderate to severe. *Fundus* is the larger part of a hollow organ that is the farthest away from the eye's opening. *Optic disc* is a small spot on the surface of the retina, called the blind spot. It is the point where the fibers of the retina leave the eye and become part of the optic nerve.

Recent surveys [13−15] show many metaheuristic search techniques have been used to detect the OD in an eye fundus image. Many researchers have proven that the differential evolution (DE) is superior to other evolutionary algorithms for many real-world applications [16−21] and, to the best of our knowledge, DE has not yet been investigated for this application. Hence, in this chapter, classical DE is applied to detect the OD.

Even though DE is highly popular, the convergence rate is very poor [22]. To improve the convergence rate, modifications have been done in the scaling factor [23], mutation rule [24], and crossover strategies [25]. In spite of its global search capability, DE has a stagnation problem due to its weak local search capability [22]. Hence, a local search is incorporated into the DE algorithm to overcome this stagnation problem. Incorporating a local search in DE is called the memetic differential evolution (MDE) [26]. Memetic algorithms (MAs) are a class of stochastic global search heuristics in which evolutionary algorithm-based approaches are combined with problem-specific solvers [22,27], approximate methods [28], or a local search [29,30]. Recent surveys indicate biomedical applications leverage the use of memetic algorithms.

The objective of this chapter is to localize the OD in an eye fundus image. To achieve this, the classical differential evolution (DE) algorithm with an intensity estimation method as a fitness function is applied to localize the OD properly. Also a local search proposed in Ref. [29] with a modification, is incorporated into the DE algorithm, that is, a MDE algorithm, to achieve a solution with more accuracy in a less number of generations. The evolutionary process of DE and MDE is analyzed based on the entire optimization process, accuracy, convergence speed, and reliability.

This chapter is organized as follows. Section 9.2 describes the implementation steps of DE. Section 9.3 explains the MDE. The experiments and results are discussed in Section 9.4. The study is concluded in Section 9.5.

9.2 CLASSICAL DIFFERENTIAL EVOLUTION FOR OPTIC DISC LOCALIZATION

DE is a population-based stochastic optimization technique introduced by Price et al. [31] and a detailed description is given in Ref. [32]. The overall flow diagram of OD localization using DE is given in Fig. 9.1. The following steps are used to implement DE for OD localization in an eye fundus image.

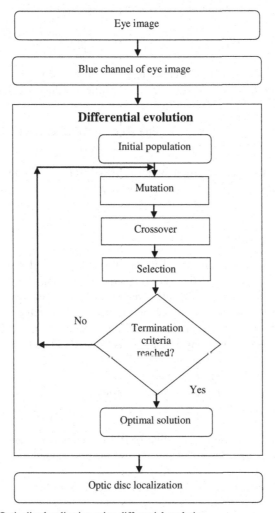

FIGURE 9.1   Optic disc localization using differential evolution.

Step 1: Generate a population of n chromosomes randomly. Each chromosome is represented by four variables i, j, $r1$, and $r2$. [12]. All these variables form a square in an image to find the OD location. Among these variables, i and j represent the center of the square, $r1$ and $r2$ determine the size of the square. The lower and upper bounds of $[i, j, r1, r2]$ are between $[150, 150, -10, -10]$ and $[1152-150, 1500-150, 10, 10]$.

Step 2: Choose any three chromosomes randomly in the current generation G and compute the mutant chromosome $v_{i,G}$ for each target chromosome $x_{i,G}$, as shown in Eq. (9.1).

$$v_{i,G} = x_{r1,G} + F\left(x_{r2,G} - x_{r3,G}\right) \tag{9.1}$$

Step 3: Perform crossover between mutant and target chromosomes to form the trial chromosome $u_{i,G}$, as shown in Eq. (9.2).

$$u_{i,G} = u_{j,i,G} = \begin{cases} v_{j,i,G} & \text{if}(\text{rand}_j(0,1) \leq C_r \text{ or } j = j_{\text{rand}}) \\ x_{j,i,G} & \text{otherwise} \end{cases} \tag{9.2}$$

Step 4: Replace the target chromosome with the trial chromosome based on the condition shown in Eq. (9.3). The fitness value is calculated as per Section 9.2.1.

$$x_{i,G+1} = \begin{cases} x_{i,G} & \text{if}\left(\text{fitness}\left(x_{i,G}\right) \leq \text{fitness}\left(u_{i,G}\right)\right) \\ u_{i,G} & \text{otherwise} \end{cases} \tag{9.3}$$

Step 5: If the end condition is not satisfied, repeat to Step 2. Otherwise, place the best chromosome in the current population.

Fitness Evaluation

The fitness function is employed to calculate the survival probability of all the chromosomes in the evolution. This fitness function, shown in Eq. (9.4), estimates the intensity of the square region in an eye fundus image and the best chromosome is selected based on the least fitness value. The procedure of calculating fitness value [12] for each chromosome $[i, j, r1, r2]$ is:

Step 1: Select the blue layer as the target image (T) and consider a square region R in T with height, $h = (r_{\text{mid}} + r1) \times 2 + 1$ and width, $w = (r_{\text{mid}} + r2) \times 2 + 1$, where $r_{\text{mid}} = 90$

Step 2: The square region is formed through:

$$R = T(i - r_{\text{mid}} - r1:i + r_{\text{mid}} + r1, j - r_{\text{mid}} - r2:j + r_{\text{mid}} + r2)$$

Step 3: Find the minimum value (m) in each row of R and fill 1 in that place of R if its value is greater than m in the corresponding row, otherwise fill in a 0.

Step 4: Calculate the fitness value as shown in Eq. (9.4)

$$y = 1000 \times \left(1 - \frac{\sum\limits_{k=1}^{h} \sum\limits_{l=1}^{w} R_{(k,l)}}{(r_{max} \times 2)^2 \times 255} \right) \tag{9.4}$$

where $r_{max} = r_{mid} + c$, c = allowed change in radius = 10

9.3 MEMETIC DIFFERENTIAL EVOLUTION

DE with a local search is called MDE. This section adopts a local search proposed in Ref. [29] with a modification suitable for this application. The local search is performed for each chromosome in the population as shown in Eq. (9.5) and the search is guided toward the best chromosome.

$$x_i' = r_1 x_i + (1 - r_1)(- (x_i - x_{best})) \tag{9.5}$$

where $x_i = i$th chromosome in the current population; x_{best} = best chromosome in the current population; r_1 = a random number between 0 and 1. The overall flow diagram of OD localization using MDE as given in Fig. 9.2.

The steps involved in MDE are:

Step 1: Generate a population of n chromosomes randomly.

Step 2: Compute the mutant chromosome $v_{i,G}$ for each target chromosome $x_{i,G}$, as shown in Eq. (9.1).

Step 3: Perform crossover between mutant and target chromosome to form trial chromosome $u_{i,G}$, as shown in Eq. (9.2).

Step 4: Replace target chromosome by trial chromosome based on the condition shown in Eq. (9.6).

$$x_{i,G} = \begin{cases} u_{i,G} & \text{if} \left(\text{fitness} \left(u_{i,G} \right) \leq \text{fitness} \left(x_{i,G} \right) \right) \\ x_{i,G} & \text{otherwise} \end{cases} \tag{9.6}$$

Step 5: For each target chromosome $x_{i,G}$, if (rand(1) < 0.5), then compute $x_{i,G}'$ as shown in Eq. (9.5)

Step 6: Replace $x_{i,G}$ by $x_{i,G}'$ based on the condition shown in Eq. (9.7).

$$x_{i,G+1} = \begin{cases} x_{i,G}' & \text{if} \left(\text{fitness} \left(x_{i,G}' \right) \leq \text{fitness} \left(x_{i,G} \right) \right) \\ x_{i,G} & \text{otherwise} \end{cases} \tag{9.7}$$

Step 7: If the end condition is not satisfied, repeat Step 2. Otherwise, return the best chromosome in the current population.

9.4 PERFORMANCE ANALYSIS

The images from the DIARETDB1 database [33] is used for this experimental purpose. This database consists of 89 images captured with 50p field of view, each of size 1152×1500 pixels, saved in PNG format. Among 89

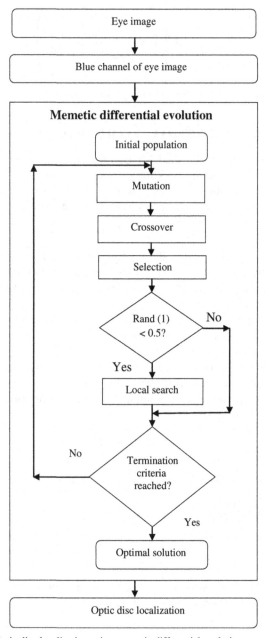

FIGURE 9.2 Optic disc localization using memetic differential evolution.

images, four images are of normal eye fundus and the remaining 84 images are of eye fundus with diabetic retinopathy. The programs are implemented in MATLAB R2008b and Intel Core i3 CPU M380 using a 2.53 GHz processor with 4 GB of RAM. The parameters for DE simulation include initial population, scale factor (F), crossover probability, and termination criteria. The recommended population size of DE simulation may vary between D and $10D$, where D represents the dimension of a chromosome. In our application, the dimension of a chromosome is 4 and, hence, the population size is set at 20. Images are chosen for this study based on their complexity levels shown by the SI_{mean} indicator [34]. Four images—image038, image001, image005, and image029—shown in Fig. 9.3, are used to find the optimal parameter values. Many experiments are carried out by changing F from 0.1 to 0.9 and crossover probability (Cr) from 0.5 to 0.9, both with an interval of 0.1. The evolution of the best fitness value of the different combinations of F and Cr is shown in Fig. 9.4, which exposes the better parameter values for DE simulation (listed in Table 9.1).

Fig. 9.5 provides the eye fundus images and their SI_{mean} values taken from DIARETDB1 database. The performance of OD detection methods is

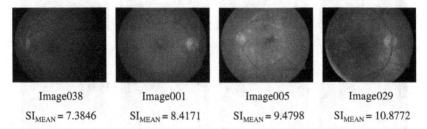

Image038	Image001	Image005	Image029
SI_{MEAN} = 7.3846	SI_{MEAN} = 8.4171	SI_{MEAN} = 9.4798	SI_{MEAN} = 10.8772

FIGURE 9.3   Images used for parameter settings.

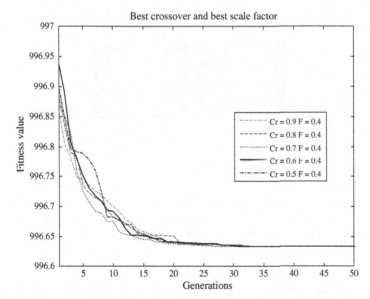

FIGURE 9.4   Best unfitness value progression of different parameter combinations for DE.

TABLE 9.1 Simulation Parameter Settings for DE and MDE

Parameter	Values
Population size	20
Scaling factor (*F*)	0.4
Crossover probability	0.7
Generations	50
Number of independent runs	20

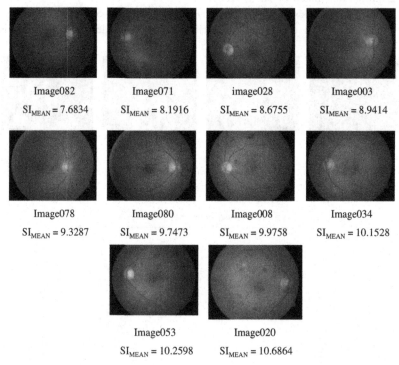

Image082   Image071   image028   Image003

$SI_{MEAN} = 7.6834$ $SI_{MEAN} = 8.1916$ $SI_{MEAN} = 8.6755$ $SI_{MEAN} = 8.9414$

Image078   Image080   Image008   Image034

$SI_{MEAN} = 9.3287$ $SI_{MEAN} = 9.7473$ $SI_{MEAN} = 9.9758$ $SI_{MEAN} = 10.1528$

Image053   Image020

$SI_{MEAN} = 10.2598$ $SI_{MEAN} = 10.6864$

FIGURE 9.5   Eye fundus images.

assessed by comparing the OD center difference between manually labeled coordinates (x_{OD}, y_{OD}) and detected coordinates (x_E, y_E) [35]. The detected OD center considers it a success if it satisfies Eq. (9.8).

$$\sqrt{(x_{OD} - x_E)^2 + (y_{OD} - y_E)^2} \leq 1R \qquad (9.8)$$

where $1R$ means 1 OD radius. For the DIARETDB1 database the maximum radius was 90 pixels in length.

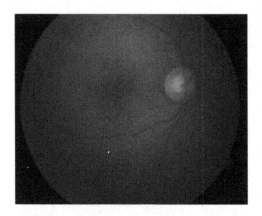

FIGURE 9.6 Detected OD in image082.

TABLE 9.2 OD Detection Results for DE and MDE

Images	DE		MDE	
	Distance Between Manually Labeled OD Center and Detected OD Center	Computation Time (s)	Distance Between Manually Labeled OD Center and Detected OD Center	Computation Time (s)
Image082	5.6569	1.8114	2.2884	3.0165
Image071	11.6619	1.7351	5.3852	3.1982
Image028	5.831	1.7517	3.6056	2.969
Image003	4.2426	1.7033	3	2.8748
Image078	3.1623	1.7189	2.2361	2.9959
Image080	3.6056	1.7109	2.8284	3.1093
Image008	5.6569	1.7534	5	2.9824
Image034	15.8114	1.7282	4.2426	2.8828
Image053	5.6569	1.7298	4	2.9986
Image020	6	1.7198	2.2361	2.9047
Average	6.72855	1.73625	3.48224	2.99322

Fig. 9.6 shows the detected OD region in an image (image082). Table 9.2 lists the computation time of DE and MDE and shows the distance between manually labeled OD center and the detected OD centers for DE and MDE. The average distance computed by DE is greater than MDE and reveals that

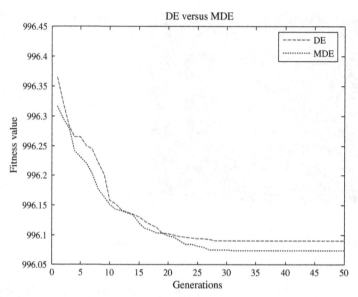

FIGURE 9.7 Best unfitness value progression of DE and MDE.

MDE can reach a close optimal solution. Although the computation time of MDE is greater than DE, it can be preferred for providing a better optimal solution. The evolution of average best chromosome fitness value of all images for both DE and MDE algorithms are shown in Fig. 9.7. From Fig. 9.7, it is visually clear that MDE is better than DE. The different performance measures shown in Table 9.3 are considered to analyze the efficacy of DE and MDE. The details of these measures are provided in [19].

Tables 9.4 and 9.5 show the $f_a(k)$ and $\overline{\text{BOG}}$ values after the 25th and 50th generations for both DE and MDE. From the Tables 9.4 and 9.5, it is clearly revealed that MDE performs better than DE over the entire optimization process. Table 9.6 gives the Acc_k value of DE and MDE which shows the accuracy value of MDE is greater than DE. MDE achieves 100% optimization accuracy, whereas DE achieves only 96.4%; however, MDE achieves it in 25 generations.

From Table 9.7, the average P measure value of MDE is 1.0 whereas for DE it is 0.925, which indicates that MDE is able to reach feasible solutions in all independent runs. AFES measure and SP measure of DE and MDE are shown in Tables 9.8 and 9.9, respectively, and the lower value is preferred for both measures. From Table 9.8, the average AFES measure value of MDE is 31.62 whereas for DE it is 39.78, which indicates that MDE is able to reach the feasible solutions in less generations than DE in all independent runs. From Table 9.9, the average SP measure value of MDE is 31.62 whereas for DE it is 45.38, which indicates that the searching performance of MDE is more successful than DE. The computation time of DE and MDE

TABLE 9.3 Performance Measures

S. No	Performance Measures	Formulae
1	Average best unfitness value, $f_a(k)$	$f_a(k) = \dfrac{\sum\limits_{runs=1}^{n} \text{Best unfitness value }(k)}{n}$
2	Average best-of-generation, $\overline{BOG}$	$\overline{BOG} = \dfrac{1}{n}\dfrac{1}{k}\sum\limits_{r=1}^{n}\sum\limits_{g=1}^{k} f(BOG_{rg})$
3	Optimization accuracy, Acc_k	$Acc_k = \dfrac{f_a(k) - Min_s}{Max_s - Min_s}$
4	Probability of convergence, P	$P = \dfrac{s}{n}$
5	Average number of function evaluations, AFES	$AFES = \dfrac{1}{s}\sum\limits_{i=1}^{s} EVAL_i$
6	Successful performance, SP	$SP = \dfrac{AFES}{P}$

where n = number of independent runs; k = generations; $f(BOG_{rg})$ = the unfitness value of the best solution at generation g of run r (among n independent runs); Min_s = worst known solution in the search space; Max_s = best known solution in the search space; s = number of successful trials/runs; $EVAL_i$ = number of evaluations required to reach the best known value in the successful trial i.

TABLE 9.4 Average Best Unfitness Value $f_a(k)$

Images	DE		MDE	
	After 25 Generations	After 50 Generations	After 25 Generations	After 50 Generations
Image082	996.0827	996.0790	996.0794	996.0790
Image071	996.1484	996.1174	996.1349	996.1162
Image028	996.0680	996.0675	996.0650	996.0648
Image003	996.0720	996.0708	996.0709	996.0708
Image078	996.0641	996.0639	996.0642	996.0639
Image080	996.0630	996.0629	996.0631	996.0629
Image008	996.0671	996.0656	996.0658	996.0655
Image034	996.2419	996.2364	996.1251	996.0806
Image053	996.0665	996.0665	996.0665	996.0665
Image020	996.0707	996.0696	996.0704	996.0696
Average	996.0944	996.0900	996.0805	996.0740

TABLE 9.5 Average Best-of-Generation $\overline{BOG}$

Images	DE		MDE	
	1−25 Generations	26−50 Generations	1−25 Generations	26−50 Generations
Image082	996.1806	996.0793	996.1232	996.0791
Image071	996.2853	996.1198	996.3267	996.1186
Image028	996.1259	996.0677	996.1021	996.0649
Image003	996.1787	996.0709	996.1428	996.0708
Image078	996.0733	996.0640	996.0886	996.0639
Image080	996.0880	996.0630	996.1020	996.0630
Image008	996.1556	996.0660	996.1537	996.0655
Image034	996.3628	996.2372	996.2645	996.0826
Image053	996.1039	996.0665	996.1043	996.0665
Image020	996.1715	996.0698	996.1562	996.0697
Average	996.1726	996.0904	996.1564	996.0745

TABLE 9.6 Optimization Accuracy Acc_k

Images	DE		MDE	
	After 25 Generations	After 50 Generations	After 25 Generations	After 50 Generations
Image082	0.9826	1.0000	0.9982	1.0000
Image071	0.9237	0.9972	0.9557	1.0000
Image028	0.9909	0.9926	0.9994	1.0000
Image003	0.9969	1.0000	0.9997	1.0000
Image078	0.9985	1.0000	0.9977	1.0000
Image080	0.9996	1.0000	0.9992	1.0000
Image008	0.9956	0.9997	0.9992	1.0000
Image034	0.6389	0.6512	0.9003	1.0000
Image053	1.0000	1.0000	1.0000	1.0000
Image020	0.9962	1.0000	0.9972	1.0000
Average	0.9523	0.9641	0.9847	1.0000

TABLE 9.7 Probability of Convergence (P)

Images	DE	MDE
Image082	1	1
Image071	0.90	1
Image028	0.85	1
Image003	1	1
Image078	1	1
Image080	1	1
Image008	0.95	1
Image034	0.55	1
Image053	1	1
Image020	1	1
Average	0.925	1

TABLE 9.8 Average Number of Function Evaluations (AFES)

Images	DE	MDE
Image082	37.95	29.05
Image071	46.17	39.75
Image028	46.94	35.05
Image003	33.65	28.05
Image078	43.50	28.55
Image080	34.10	30.90
Image008	45.32	31.65
Image034	49.18	36.55
Image053	23.70	22.90
Image020	37.30	33.70
Average	39.78	31.62

is 0.035 s and 0.059 s, respectively, on average to complete a generation. Although MDE takes more time than DE to complete a generation, MDE achieves the solution in a smaller number of generations with 100% accuracy, whereas DE achieves it in a higher number of generations with only 96.4% accuracy.

TABLE 9.9 Successful Performance (SP)

Images	DE	MDE
Image082	37.95	29.05
Image071	51.30	39.75
Image028	55.22	35.05
Image003	33.65	28.05
Image078	43.50	28.55
Image080	34.10	30.90
Image008	47.70	31.65
Image034	89.42	36.55
Image053	23.70	22.90
Image020	37.30	33.70
Average	45.38	31.62

TABLE 9.10 One Tailed t-Test Results Between DE and MDE

Measures	P Value in t-Test	Significance Level
Average unfitness value	0.0223	0.05
Average best of generations	0.003	
Optimization accuracy	0.0163	

The values of P measure, AFES measure, and SP measure clearly show the distinct difference between MDE and DE, which proves that the results of MDE are better than those of DE. In order to strengthen the inference from the empirical study of $f_a(k)$, $\overline{BOG}$ and Acc_k, the one tailed t-test is carried out to statistically compare the performance of MDE and DE. One tailed t-test is a hypothesis test in which the critical area of interest falls in just one tail of a T distribution. It is performed when the results are attracted to a particular direction. Therefore at the 5% significance level the assumptions made are H_0 (null hypothesis): there is no significant difference between the MDE and DE, H_1 (alternative hypothesis): MDE is better than DE. From Table 9.10, the P-value for the considered performance measures are less than 0.05, which shows the rejection of the null hypothesis and indicates that MDE is better than DE.

9.5 CONCLUSION

DE has been applied to localize the OD in eye fundus images. A local search is incorporated in DE to improve the localization accuracy of the OD. The proposed MDE is able to produce close optimum results. Integrated local search using DE algorithms enhances the search capability and reduces the number of generations to achieve the solution. In addition, the performance assessment criteria, such as accuracy, entire optimization process, convergence speed, and reliability, has been taken into consideration for comparing DE and MDE results. The analysis report reveals that MDE guarantees an optimal solution in less generations and the same is confirmed by using a statistical hypothesis test. A possible direction for future research can be to investigate methods to reduce the distance between the manually labeled OD center and the detected OD center further.

REFERENCES

[1] D. Amin, A. Hamid, M. Mohammad-Shahram, Optic disc localization in retinal images using histogram matching, EURASIP J. Image Video Process. 19 (2012) 1−11.

[2] P. Nagarajan, S.S. Vinsley, Optic disc boundary approximation using elliptical template matching, Int. J. Pharm. Sci. Rev. Res. 39 (2) (2016) 270−275.

[3] R. Kamble, M. Kokare, G. Deshmukh, F.A. Hussin, F. Mériaudeau, Localization of optic disc and fovea in retinal images using intensity based line scanning analysis, Comput. Biol. Med. 87 (2017) 382−396.

[4] V. Kumar, N. Sinha. Automatic optic disc segmentation using maximum intensity variation, in: Proc. IEEE 2013 Tencon, Spring, Sydney, Australia, 2013.

[5] A. Gopalakrishnan, A. Almazroa, K. Raahemifar, V. Lakshminarayanan, Optic disc segmentation using circular hough transform and curve fitting, in: 2nd International Conference on Opto-Electronics and Applied Optics (IEM OPTRONIX), IEEE, Vancouver, BC, Canada, 2015.

[6] A. Fraga, N. Barreira, M. Ortega, M.G. Penedo, M.J. Carreira. Precise segmentation of the optic disc in retinal fundus images, in: International Conference on Computer Aided Systems Theory, 2012, pp. 584−591. EUROCAST

[7] A. Jayanthiladevi, A human computer interfacing application, Int. J. Pharma Bio Sci. 6 (2016).

[8] M. Lalonde, M. Beaulieu, L. Gagnon, Fast and robust optic disc detection using pyramidal decomposition and Hausdorff-based template matching, IEEE Trans. Med. Imaging 20 (11) (2001) 1193−1200.

[9] A.P. Gopi, M.S. Anjali, S. Issac Niwas, PCA-based localization approach for segmentation of optic disc, Int. J. Comput. Assist. Radiol. Surg. 12 (12) (2017) 2195−2204.

[10] P.H. Princye, V. Vijayakumari, Retinal disease diagnosis by morphological feature extraction and SVM classification of retinal blood vessels, Biomed. Res. (2018).

[11] R. Arnay, F. Fumero, J. Sigut, Ant colony optimization-based method for optic cup segmentation in retinal images, Appl. Soft Comput. 52 (2017) 409−417.

[12] G. Ferdic, S. Santhosh Baboo, GA based automatic optic disc detection from fundus image using blue channel and green channel information, Int. J. Comput. Appl. 69 (2) (2013) 23−31.

[13] S. Bharkad, Automatic segmentation of optic disk in retinal images, Biomed. Signal Process. Control 31 (2017) 483–498.

[14] N. Kulkarni, J. Amudha, A study on current optic disc detection methods, IJCT A 9 (40) (2016) 441–452.

[15] B. Vinoth Kumar, K. Janani, M. Priya, A survey on automatic detection of hard exudates in diabetic retinopathy, in: IEEE International Conference on Inventive Systems and Control, January 19–20, IEEE, 2017.

[16] S. Das, S. Sil, Kernel-induced fuzzy clustering of image pixels with an improved differential evolution algorithm, Inf. Sci. 180 (8) (2010) 1237–1256.

[17] S. Dasgupta, A. Biswas, S. Das, A. Abraham, Modeling and analysis of the population dynamics of differential evolution algorithm, AI communications, Eur. J. Artif. Intell. 221 (1) (2009) 1–20.

[18] A. Mandal, S. Das, A. Abraham, A differential evolution based Memetic Algorithm for workload optimization in power generation plants, in: 2011 11th International Conference on Hybrid Intelligent Systems (HIS), December 5–8, vol. 1 (2012), pp. 1–10.

[19] B. Vinoth Kumar, G.R. Karpagam, Differential evolution versus genetic algorithm in optimizing the quantization table for JPEG baseline algorithm, Int. J. Adv. Intell. Paradig. 7 (2) (2015) 111–135.

[20] B. Vinoth Kumar, G.R. Karpagam, Reduction of computation time in differential evolution based quantization table optimization for the JPEG baseline algorithm, Int. J. Comput. Syst. Eng. 4 (1) (2018) 58–65.

[21] B. Vinoth Kumar, G.R. Karpagam, Single versus multiple trial vectors in classical differential evolution for optimizing the quantization table in JPEG baseline algorithm, ICTACT J. Soft Comput. 7 (4) (2017) 1510–1516.

[22] B. Vinoth Kumar, G.R. Karpagam, Knowledge based differential evolution approach to quantization table generation for the JPEG baseline algorithm, Int. J. Adv. Intell. Paradig. 8 (1) (2016) 20–41. Inderscience Publishers.

[23] H. Huang, P. Hu, Self-adaptive differential evolution algorithm for the optimization design of pressure vessel, in: 2016 12th World Congress on Intelligent Control and Automation (WCICA), June 12–15, 2016, IEEE, Guilin, China.

[24] W.M. Ali, Z.S. Hegazy, T. Abd-Elaziz, Real parameter optimization by an effective differential evolution algorithm, Egypt. Inform. J. 14 (1) (2013) 37–53.

[25] K. Sandeep, K.S. Vivek, K. Rajani, Memetic search in differential evolution algorithm, Int. J. Comput. Appl. 90 (6) (2014) 40–47.

[26] D. Jia, G. Zheng, M.K. Khan, An effective memetic differential evolution algorithm based on chaotic local search, Inform. Sci. 181 (2011) 3175–3187.

[27] B. Vinoth Kumar, G.R. Karpagam, Knowledge based genetic algorithm approach to quantization table generation for the JPEG baseline algorithm, Turk. J. Electr. Eng. Comput. Sci. 24 (3) (2016) 1615–1635.

[28] B. Vinoth Kumar, G.R. Karpagam, A problem approximation surrogate model (PASM) for fitness approximation in optimizing the quantization table for the JPEG baseline algorithm, Turk. J. Electr. Eng. Comput. Sci. 24 (6) (2016) 4623–4636.

[29] J. Gu, G. Gu, Differential evolution with a local search operator, in 2010 2nd International Asia Conference on Informatics in Control, Automation and Robotics, March 6–7, vol. 3 (2010), pp. 480–483.

[30] L. Peng, Y. Zhang, G. Dai, M. Wang, Memetic differential evolution with an improved contraction criterion, Comput. Intell. Neurosci. (2017) 1–12.

[31] K. Price, R. Storn, J. Lampinen, Differential Evolution: A Practical Approach to Global Optimization, first ed., Springer-Verlag, New York, 2005.

[32] S. Das, P.N. Suganthan, Differential evolution—a survey of the state-of-the-art, IEEE Trans. Evol. Comput. 15 (1) (2011) 4−31.

[33] T. Kauppi, V. Kalesnykiene, J.K. Kamarainen, L. Lensu, I. Sorri, A. Raninen, et al., DIARETDB1 diabetic retinopathy database and evaluation protocol, Med. Image Underst. Anal. 2007 (2007) 61. Available from: https://doi.org/10.5244/C.21.15.

[34] Y. Honghai, W. Stefen, Image complexity and spatial information', in: 2013 Fifth International Workshop on Quality of Multimedia Experience (QoMEX), July 3−5, (2013), IEEE, Austria, 2013, pp. 12−17.

[35] R. Veras, F. Medeiros, L.D. Santos, F. Sousa, A comparative study of optic disc detection methods on five public available database, XIV Workshop on Medical Informatics, IEEE, 2014, pp. 1772−1781.

Chapter 10

Classification of Myocardial Ischemia in Delayed Contrast Enhancement Using Machine Learning

R. Merjulah and J. Chandra

Christ University, Bangalore, India

Chapter Outline

10.1 INTRODUCTION

10.1.1 Clinical Study

Delayed contrast enhancement magnetic resonance imaging (MRI) of ischemia of myocardial disease has become a significant application in cardiovascular MRI. Myocardial infarction is the medical term for heart attack defined as myocardial cell death. Heart attack is a life-threatening illness that ensues when the flow of blood to the heart muscle is cut off, causing tissue damage in the heart muscle. A blockage can develop due to fat, cholesterol, and cellular waste products [1]. The classic experience of a heart attack is acute pain in the chest and shortness of breath. But the symptoms are quite varied. Pain in the chest is the most commonly reported symptom for both men and the women. The heart is the central organ in the cardiovascular system that includes different blood vessels of which arteries are the most important

Intelligent Data Analysis for Biomedical Applications. DOI: https://doi.org/10.1016/B978-0-12-815553-0.00011-2

blood vessels. The arteries take the oxygen-rich blood through the body and all the other organs. When these arteries are blocked or narrowed due to the buildup of fat, and the blood flow toward the heart stops completely, the blockage leads to a heart attack. Heart attack patients will be diagnosed by a doctor by listening to the heart for irregularities and checking the heartbeat. Doctors will run several tests if they suspect the patient had a heart attack. An electrocardiogram (ECG) will be included to measure the electrical activity [2]. A blood test (troponin), stress test, angiogram, and echocardiogram etc., are various other tests for suspected heart attack patients. Heart attack requires immediate treatment. The recovery chances from a heart attack depends on the damage to the heart and how quickly emergency care is given to the patient. Survival is based on the treatment received as early as possible. However, if there is major damage to the heart muscle, the heart cannot pump the required amount of blood all over the body, which leads to heart failure. Heart damage increases the risk of abnormal heart rhythms or arrhythmia. Many heart arrhythmia are harmless. Recording damage to the heart is through the visualization process. The visualization process uses computerized tomography (CT) or MRI. The MRI scanner uses radiowaves and a powerful magnet which is linked with a computer to create a remarkably clear and complete crosssection images of the body.

MRI plays an important role in the early detection and prediction of arrhythmia in evaluating the various stages of a MI. MRI has various advantages compared to other imaging modalities. To improve the quality and for the identification of the scar regions in a MI patient's MRI, the radiologist inserts a gadolinium contrast media, otherwise known as MRI contrast media, dyes, or agents—the chemical substances which are inserted in the patient's body through injection to improve the quality of the image and for clear identification of the scar regions. The patient is not required to be exposed to radiation. The spatiotemporal resolution, high contrast resolution, of the image helps visualize tissue characteristics, etc. Delayed myocardial enhancement MRI is a valuable and nonspecific imaging technique used to identify scar tissues in the heart muscle.

Delayed enhancement MR imaging is used for the diagnosis of the scar tissue in the heart (Fig. 10.1). In MI, biomedical and functional abnormality starts immediately after the inception of ischemia. Loss of contractility occurs within 60 seconds and the loss of viability occurs in 20–40 minutes. The changes initially decrease the perfusion, leading to a weakening of tissue oxygenation, diastolic dysfunction, and systolic dysfunction, etc., finally causing tissue necrosis. Each stage is based on time. At 15 minutes, there will be ischemia myocardium, but no infarct. At 40 minutes subendocardial infarction can be seen. At 3 hours the subendocardial infarction is seen to extend to the midmyocardial region. Beyond 6 hours the infarct become transmural [3].

Fig. 10.2 illustrates the delayed enhancement of myocardial tissue. The clinical use of the delayed enhancement of myocardial tissue is commonly

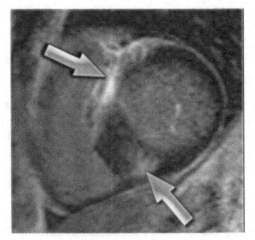

FIGURE 10.1   Short axis delayed enhancement MRI.

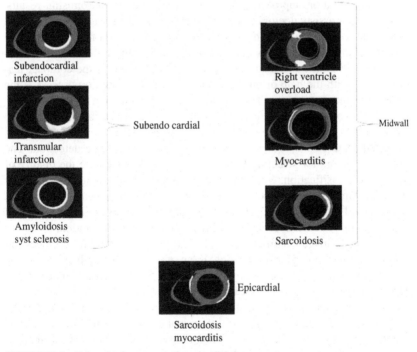

FIGURE 10.2   Delayed enhancement of myocardial tissue.

performed for the evaluation of risk in the myocardial. In the case of the ischemia, cardiomyopathy is used to evaluate the scar in the heart. The computer-aided machine is used to identify and examine ischemia cardiomyopathy within a short period of time.

10.1.2 Functional Considerations

Medical image segmentation has automatic or semiautomatic detection of the two-dimensional (2D), or three-dimensional (3D), image. Image segmentation is the procedure of dividing a digital image into a multiple set of pixels. The prior goal of the segmentation is to make things simpler and transform the representation of medical images into a meaningful subject. Segmentation is a difficult task because of the high variability in the images [4]. Furthermore, many variant modalities, such as CT, X-ray, MRI, microscopy, positron emission tomography, single photon emission computer tomography, among others, makes segmentation difficult. The challenging problem is for segmenting the regions with missing edges, absence of texture contrast, region of interest (ROI), and background. To report these issues, many segmentation approaches have been proposed with promising results. Segmentation assists doctors to diagnose and make decisions. After segmentation, the defected features have to be extracted through a feature extraction process. Feature extraction is the type of dimensionality reduction that effectively represents the defected region of a medical image as the compact feature vector with the help of the ROI [5]. The feature extraction process is the attribute reduction process. The feature is the representation of the combination of keywords. The feature detection and extraction are combined to solve the computer vision problem. Once the features have been extracted, the extracted features will be used to build models for accurate detection. The important factors for selecting features extraction will increase the quality of the image classification. So, human expertise is often essential to translate raw data into the set of useful features with the help of feature extraction algorithms. Different classifications of feature extractions are compression of data, decomposition and projection of data, and pattern recognition.

In general, image classification is the next process in the image processing system. Medical image classification is an important research arena in the developing attention of the research community and the medical industry [6]. The objective of image classification is, perhaps, the most important part of digital image analysis. The intent of the classification procedure is to sort all the pixels in a digital image into one of several classes. Manual image classification is not always reliable and is time consuming. In biomedical applications, the automatic technique of classification could help large-scale image datasets to promote faster diagnosis [7]. The image classification methods are the clinical diagnosis tools based on the medical images. In classification, a class represents the part of the body and tissue from the organ. Image data is has enormous practical significance in medical information.

In this chapter, the MI short axis delayed enhancement of MRI is considered for the classification of MI. In MI images, the scar regions are detected through the segmentation process. Using FCM multispectral and single channel, the basic idea is to segment the scar tissue in the MI patient's MRI which partitions the image into hard and fuzzy clustering techniques. The feature extraction process is completed through the morphological filtering technique. Further on, for classification of the MI, the data is passed through the feed forward neural network (FFNN) using Levenberg–Marquardt Back Propagation (LMBP). However, in the medical field there has been issues like privacy, security, and the analysis of the data. Additionally, there are other challenges like detecting, classifying, and diagnosis of the MI. To solve these complex problems, the intelligence technique can assist the doctor as a secondary opinion. The main motivation of this chapter is to detect and classify the MI with the help of the proposed method by using MRI. Research studies have been dedicated to the processing and analyzing of the MI dataset to extract meaningful information, such as detecting and classifying cardiac MI.

10.2   RELATED WORK

Tantimongcolwat et al., [8] proposed a machine-learning technique, the Back Propagation Neural Network (BNN) and Direct Kernel Self Organizing Map, to interpret the Ischemia Heart Disease (IHD) pattern recorded by magneto-cardiography (MCG) for saving the manual interpretation time. 125 datasets divided into 75 cases of training sets and 51 cases of testing sets produces 89.7% for sensitivity, 54.5% for specificity, and 74.5% accuracy for BNN, and 86.2% for sensitivity, 72.7% for specificity and 80.4% accuracy for MCG, respectively.

To achieve strong clusters Kannan et al. [9] proposed an objective function to obtain the new hyper tangent function from the exited hyper tangent function to perform efficiently on a large dataset with more noisy medical images. An artificially generated dataset is used for the experiments to show the efficiency of the FCM to implement the proposed methods for the segmentation of, for example, breast medical images. The classification of the proposed FCM segmentation method is obtained using the Silhouette method.

Nanthagopal et al. [10] proposed the support vector machine (SVM)-based classification on dual intensity extraction based on the features followed by principal component analysis (PCA) processed for feature selection and achieves an accuracy rate of 96%.

He et al. [11] proposed improved FCM with a constraints algorithm for the segmentation of brain tissue based on diffusion tensor imaging (DTI). To overcome the problem of the FCM algorithm, the proposed FCM is used to change the membership function and objective function of the standard FCM for the enhancement of the noisy immunity. The proposed algorithm is based on the multichannel feature of DTI for tissue segmentation.

To improve the efficiency of the conventional FCM algorithm, Adhikari et al. [12] proposed the spatial Fuzzy C-Means (scFCM) algorithm. Adding both the global and local spatial information into the weighted membership function is to decrease the sensitivity problem to the noise and intensity inhomogeneity in MRI data. The csFCM algorithm has achieved higher accuracy in terms of tissue segmentation accuracy, segmentation accuracy, cluster validation functions, and the receive operating characteristic (ROC) curve.

For assessing the severity of nuclear cataracts from slit-lamp images, Gao et al. [13] proposed an automatic system to learn the features. Local filters are trained through image patches applied into a CNN with a group of recursive neural networks for the extraction of high-order features. To determine the cataract grade, support vector regression is included. The proposed model was validated using 5378 images which out performs state-of-art models with an error rate of 0.322, integral agreement ratio of 68.6%, a decimal grading error of 86.5% with ≤ 0.5, while the decimal grading error is 99.1% with ≤ 1.0.

To find out the disease status of a patient, Payan et al. [14] proposed a deep-learning method with sparse auto encoders and 3D CNN. The result outperforms the other classifications. The experiment was tested with the ADNI dataset based on MRI brain images.

Shen et al. [15] proposed a hierarchical learning framework with the multiscale CNN for capturing nodule patches for the diagnosis of lung nodule classification without the definition of nodule morphology to learn a set of class-specific features to quantify nodule characteristics. The method was evaluated using the CT images.

Antony et al. [16] proposed automated Deep CNN to find the severity in knee osteoarthritis (OA). The deep CNN model is trained with ImageNet with fine tuning on knee images of OA. The X-ray dataset images with the Deep CNN shows a substantial improvement over the current state-of-art techniques.

Asl et al. [17] proposed deep 3D-CNN to predict Alzheimer's disease (AD), which learns the generic features capturing biomarkers on AD and different domain dataset adaption. To capture the anatomical shape variation in the structure of brain MRIs, the 3D-CNN is developed on a 3D convolutional auto encoder. The fine tuning takes place in the fully connected upper layers of the 3D-CNN for every specific classification of AD. The experimental work was done on the ADNI MRI dataset.

To classify skin lesions, Kawahar et al. [18] proposed CNN architecture which learns based on the evidence from multiple image resolutions while leveraging pretrained CNN. To analyze the image at various resolutions along with the learning interaction across multiple image resolutions for the same view of the field, the CNN is trained and composes multiple tracts. The classification accuracy outperforms other state-of-art multiscale approaches.

Choudhry and Kapoor et al. [19] compared various FCM variations for the segmentation in cluster validity function, feature structure, and fuzzy partition based on the performance of MRI noisy images in terms of two classes. The compared methods are FCM-S1, FCM-S2, TEFCM, BCFCM, DFCM, SFCM, FLICM, MDFCM, TEFCM, RFCMK, WIPFCM, and KWFLICM. Analysis is based on the partition entropy, partition coefficient, time complexity, and segmentation accuracy. The comparative result reveals that FLICM gives better accuracy on the noisy images followed by WIPFCM, MDFCM, and PFCM.

Cao et al. [20] proposed the Map Reduce Parallel Programming model that changes the basic idea of the Adaboost-Back Propogation (BP) neural network. The research work starts by constructing a strong classifier based on the Adaboost algorithm and then by using the parallel Adaboost−BP neural network and the feature extraction algorithm they designed maps and reduces tasks. Finally, using a Hadoop cluster the researchers developed an automated classification model. The training and testing data for the classification was done using the Pascal VOC 2007 and Caltech 256 dataset. The result reveals an increase in the classification accuracy by 14.5% and 26.0% and the computational time than the traditional Adaboost BP neural network.

To segment the nontrivial task of lip segmentation, Danielis et al. [21] proposed Lambertian illumination for face image enhancement, lip segmentation using morphological operation, and, finally, Fourier-based modeled lip boundaries for feature extraction. The result reveals the effectiveness through low spectral range, noise, and moustache.

The feature of Kernel fuzzy clustering along with the edge-based active contour method with the distance regularized level set and spatial constraints was proposed by Gupta and Anand [22] for the correct segmentation of ultrasound medical images. The experiment was tested with ultrasound images and synthetic images. The proposed segmentation provides better performance in terms of computational processing speed.

Yang [23] proposed the progressive support-pixel correlation statistical method for medical images by categorizing the experimental data as actual single-spectral mammograms, computer-simulated images, and MRI multispectral breast images.

In the conventional Hard C-Means (HCM) algorithm, Gharieb et al. [24] proposed a hybrid of local spatial membership functions to spatially smooth the pixel vicinity and the data information is incorporated in a Fuzzy clustering technique for image segmentation. The divergence between the pixel membership function and the smoothed one is added to the HCM objective function for fuzzification. By adding a weighted HCM function the resulting fuzzified HCM is regularized. The proposed algorithm has the accuracy of 92%, sensitivity is 84%, and specificity is 94.7%.

Pei et al. [25] proposed a density-based Fuzzy C-Means (D-FCM) algorithm by introducing the density of each sample. The clusters' number and the initial membership matrix are defined by the density peaks automatically.

Dohara et al. [26] proposed SVM to detect myocardial infarction using 12 lead ECG system. The myocardial infarction detection sensitivity is 96.66%, specificity is 100%, and the accuracy is 98.33% for the original features of 220 images. The proposed research applies the PCA reduction technique for the reduction from 220 to 14 features and achieves 96.66% for sensitivity, 96.66% for specificity, and accuracy of 96.66%.

Guo et al. [27] proposed the Fuzzy feature fusion method for the automatic segmentation of gliomas in radiotherapy using MRI, which includes fractional anisotropy, relative cerebral blood volume, and apparent diffusion coefficient. To transform the volume of the image, every functional modality of the histogram-based fuzzy model is created. Tumors were generated automatically based on the fusion result of the fuzzy feature regions and spaces. Comparative results of manual delineated and auto-segmented gross tumor volume shows that the proposed method can reach high accuracy.

A thresholding extraction method grounded on variational mode decomposition has been proposed by Li et al. [28]. The experimental results shows that the minimum point search scheme is more efficient than the cross point search method. The result is compared with the FCM, particle swarm optimization algorithm, and bacterial foraging algorithm. The result gives a similar performance for all the compared algorithms, but the proposed algorithm is faster in computation speed.

Devkota et al. [29] proposed the mathematical morphological reconstruction to detect and identify brain tumors. It is an early detection technique that removes the noise and artifacts and it is segmented to find the ROI.

The improved FCM clustering algorithm allows the segmentation of the images and to correct the intensity inhomogeneity [30]. Generalized FCM methods for the modified fast FCM algorithm [31,32], Rough set-based FCM algorithm [11,33,34], FCM-based algorithm [9,11,35] for the brain segmentation, Fuzzy set-based clustering [36] approaches for brain image segmentation.

Traditionally, different machine-learning algorithms, including K-nearest neighbor, support vector machine, and deep CNN, etc., are there for the classification of medical images. However, image classification is still a difficult problem in the field of classifying medical images. Generally, many researchers are working with image classification with the help of machine intelligence to produce optimized results in the field of image analysis. From the discussion here, it is observed that the intelligence technique for the segmentation and classification of various medical images raises accuracy.

Phase I

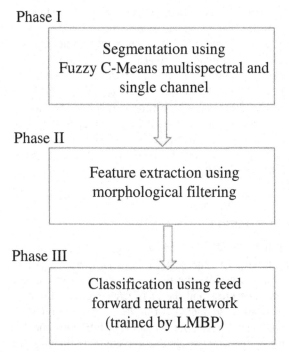

FIGURE 10.3   Proposed framework for the classification of MI.

10.3 METHODOLOGY

Current research focuses on the classification of the MI in the short axis delayed enhancement MRI. Phase I is processed through the segmentation process with the help of FCM of multispectral and single channel for the identification of the scar tissue in the heart of a MI patient's MRI. Phase II includes the process of feature extraction using the morphological filtering technique. Phase III processes the data for the classification of the MI in the delayed enhancement images using the FFNN which is trained through the LMBP algorithm. LMBP and SCG is used for training the network, and the results are compared.

Selecting the most suitable methodology for medical data is a tricky process. Each methodology has advantages and disadvantages. The proposed framework is defined in Fig. 10.3 that shows its efficient classification for the MI.

Fig. 10.3 shows the proposed framework for MI classification of delayed enhancement in MRI. In Phase I the segmentation takes place through the FCM multispectral and single channel for the identification of the scar tissue in the MI patient's MRI. Phase II pushes the dataset to the feature extraction using morphological filtering technique. Finally, FFNN is used for the classification of the MI data which is trained through the LMBP. The purpose of the proposed framework is to bring about high-performance accuracy on the classification of MI. The provided framework is able to classify the MI with the accuracy rate of 99.9% when it is repeatedly trained by the LMBP in the FFNN classifier. The network training for classification is processed through the LMBP algorithm and SCG for a comparative result. The comparative result reveals that training the FFNN through the LMBP algorithm yields high accuracy and the computation time is reduced. The training is repeated till it reaches high accuracy. The repeated training in the FFNN will always yield the best accuracy performance. The result shows that the proposed method is one of the best methods available for the classification of delayed enhancement MI using MRI.

10.3.1 Segmentation

Segmentation involves division of the images into the same classes or type. For example, to segment the scarred and non-scarred heart tissue, the segmentation technique divides the tissue into the distinct classes. The affected region falls into one class and the unaffected region falls into another class. There are two types of segmentation in the fuzzy model. One is hard and the other is soft or fuzzy. Assigning a pixel that falls into one class is the hard segmentation. In medical images, complete classification is not possible because the tissues contribute to various pixels. In the location of the object's boundaries the soft fuzzy part permits uncertainty. Membership function occurs for each class at every location where pixels exist in the soft segmentation. The membership function value is 0 if the pixel does not belong to the class. The membership function value is 1 if the pixel belongs to the class. The value of the membership function varies from 0 and 1. Membership on the soft task reflects the value of the pixel data and the centroid value. The class membership function approaches unity if the data value is closer to the centroid class [37].

The FCM is an unsupervised method. The algorithm works by allocating membership to each data point with respect to every cluster center on the basis of the distance between the data points and the cluster center. If the cluster point is near to the data then it belongs to the membership of the particular cluster. The summation of the membership of every data point should be 1.

After every membership iteration the cluster centers are updated through:

$$\mu_{ij} = \frac{1}{\sum_{k=1}^{c} \left(\frac{d_{ij}}{d_{ik}}\right)^{\left(\frac{2}{m-1}\right)}} \qquad (10.1)$$

$$v_j = \left(\frac{\sum\limits_{i=1}^{n} (\mu_{ij})^m x_i}{\sum\limits_{i=1}^{n} (\mu_{ij})^m} \right), \forall_j = 1, 2, 3 \ldots \ldots .c \tag{10.2}$$

In Eqs. (10.1 and 10.2), the number of data points in the dataset is n, v_j represents the jth cluster center of the MI images, m is fuzziness index $\sum[1, \infty]$, The number of cluster canter represented in the MI images is c, the membership of the jth data to jth cluster center is μ_{ij}, the Euclidean distance between ith data and jth cluster center is d_{ij}.

The main objective of the FCM is to minimize the sum over the pixel using the following

$$j(u, v) = \sum_{i=1}^{n} \sum_{j=1}^{c} (\mu_{ij})^m ||xi - vj||^2 \tag{10.3}$$

In Eq. (10.3), where the Euclidean distance between the jth data and jth cluster center is $||xi - vj||$.

Steps for the Fuzzy in the MI dataset

Let $x = \{x1, x2, x3 \ldots xn\}$ is the set data points in the images
$v = \{v1, v2, v3 \ldots vn\}$ is the set of centers

Step 1: Randomly, select 'c' cluster centers for the image.

The centroid value of the MI is Centroid $1 = 95.25$, Centroid $2 = 190.5$, Centroid $3 = 285.75$ and the desired classes for the fuzzy is 3.

Step 2: Calculated the fuzzy membership using the following equation

$$\mu_{ij} = \frac{1}{\sum\limits_{k=1}^{c} \left(\frac{d_{ij}}{d_{ik}} \right)^{\left(\frac{2}{m} - 1 \right)}} \tag{10.4}$$

Step 3: Calculated the fuzzy centers using the following equation

$$v_j = \left(\frac{\sum\limits_{i=1}^{n} (\mu_{ij})^m xi}{\sum\limits_{i=1}^{n} (\mu_{ij})^m} \right), \forall j = 1, 2, 3, \ldots c, \tag{10.5}$$

Step 4: Repeated Step 2 and Step 3 till the minimum j value is achieved.

$$||u^{(k+1)} - u^{(k)}|| < \beta \tag{10.6}$$

The iteration step is K, where $k = 200$ for the MI dataset, β is the termination between [0,1], $u = (\mu_{ij})_{n*c}$ is the fuzzy membership matrix, and J is the objective function.

The advantages of the FCM is that it works without the specified set of data for describing the parameters of an algorithm. The algorithm allows the fuzzy segmentation which specifies the K-means algorithm. The clustering is based on the fuzzy membership function and the mean value, which is estimated based on each tissue class.

10.3.2   Feature Extraction

Morphology is the image processing operation that processes the images based on shape [38]. Numerous imperfections will be present in the binary images. Specifically, the binary image region produced by simple thresholding are filled with noise. By the structure and form of the image, the morphological filtering pursues the goals of removing the noise. It is a nonlinear operation based on the shape or the feature of the images. The morphological filtering operation relies on the value of the pixel, but not on numerical value. The operations are best-suited for binary images, but can also be applied to grayscale images. A small shape or template of a structural element will be yielded by the morphological technique. At every possible location of an image the structuring element is positioned and related with the matching neighborhood pixels. Specific operations will be tested whether the elements fit with the neighborhood, while the other operations test whether it hits or intersects the neighborhood.

Every pixel is connected with the neighboring pixel within the structuring element, when the structuring element is on the binary image. The structuring element is to suitably fit the image if the pixel is set to 1 and the corresponding pixel image is also 1. Similarly, the structuring element is to hit or intersect an image unless and until one of the pixels set to 1 and the equivalent pixel image is also 1. 0-valued pixels of the structuring element are ignored as it indicates the points where the equivalent image value is irrelevant.

The purpose of morphological processing is primarily to remove imperfections during segmentation. The basic operations are dilation and erosion. Dilation improves pixels to the boundaries and erosion eliminates pixels on object boundaries. The basic operation is the shift invariance.

Since the dataset is the grayscale image, the grayscale image erosion of f by b is:

$$(A\theta B)(x) = \inf_{y \in B} [f(x+y) - b(y)] \tag{10.7}$$

In Eq. (10.7) inf is the infimum, x and y values are the filter size where $x = 5$ and $y = 5$ for the MI dataset. The minimum of points in the neighborhood is the erosion point, which is defined by the structuring element. It is similar to median and Gaussian filtering.

10.3.3   Classification

Neural network (NN) contains several hidden layers. Each hidden layer in the NN consists of multiple nodes. Each node is connected to the input layer connections and output layer connections. Every connection has a different weight, which is adjustable. Data is passed through these hidden layers with different results. Each layer represents a parameter from the dataset. Before the data insertion into the NN, the dataset has to be preprocessed and normalized. The individual nodes in the NN exit in the neurons with the help of the input data and performs the simple elements operating in parallel. The elements are inspired by biological systems. The elements are connected together to form the network function. Weight values which are associated with each node is determined by the iterative flow of the training data. The training of a neural network for a particular function occurs by changing the weights between elements, so the particular input points to the specific output. Once the NN is trained, it can be applied to the new data for classification by the trained network which is activated through the NN by passing the relevant data sources and then the forward flow of the data through the network and the activation of the output nodes. Based on the comparisons of the target on the output, the network is adjusted till the output network matches the target [39]. The output nodes show each pixel's classification.

Steps for the proposed FFNN:

Step 1: Input a set of trained data.
Step 2: For the training of x, set the corresponding input activation $a^{x,1}$ and perform the following steps:
- Feed Forward: For each $I = 2, 3, 4, \ldots L$ compute:

$$z^{x,1} = W^l a^{x,l-1} + b^l \text{ and } a^{x,1} = \sigma(Z^{x,l})$$

Output error $\delta^{x,L}$ computes the vector

$$\delta^{x,L} = \nabla_a c_x \odot \sigma^1(z^{x,L})$$

- Back propagate the error using the following derivation

$I = I, -1, L - 2, \ldots 2$ compute

$$\delta^{x,I} = ((W^{l=1})^T f^{x,l+1}) \odot \sigma^1(z^{x,l})$$

Step 3: Gradient Descent
For each $I = L, L - 1, \ldots 2$ update the weights according to the rule $W^l \rightarrow W^l - \frac{n}{m} \sum_x \delta^{x,l}(a^{x,l-1})^T$ and the biases according to the rule $b^l \rightarrow b^l - \frac{n}{m} \sum_x \delta x^l$

The proposed FFNN is a two-layered network with sigmoid hidden neurons and linear output neurons. The network is trained using the LMBP algorithm. Training data changes according to its errors. Validation is used for the measurement of the network generalization and paused when the generalization stops. Testing had no effect on training and provides an independent measure of network performance during and after training. The proposed FFNN has been trained often to perform the complex functions in the field of classification. Most widely, the FFNN algorithm trained using the LMBP is used for image classification.

10.3.4 Training Feed Forward Neural Network

FFNN belong to the artificial neural network (ANN) family, where connections between the units do not form a circle. The single-layer perceptron network is a simpler form of the FFNN. The input values are fed to the outputs through a series of weights. So, it is considered as the simplest kind of the FFNN. Back propagation calculates the gradient error of the network along with the networks' modifiable weights. To reduce the weight, the gradient is always used in a simple gradient decent algorithm. It is a multilayered feed forward network with one hidden unit. The architecture of the back propagation is similar to the multilayered feed forward network. The time taken for computation will increase if there are more hidden layers. Based on the back propagation, the LMPB training algorithm is proposed.

Four stages are required for the training of the back propagation:

1. Random values are initialized as the weight.
2. In each input units 'X' receives an input signal and transmits to the hidden units $y1, y2, y3, y4 \ldots yn$. Activation function is calculated in the hidden layers and send the signal to output units.
3. In back propagation of error.
4. Changes of the weights and biases.

Technically, for training the weights in multilayers, the back propagation algorithm is suggested. The structure of the network also has to be defined of one layer connected to the other layer. One input layer, one hidden layer, and an output layer is the standard network structure. Back propagation can be used both in regression as well as classification problems.

10.3.5 Levenberg–Marquardt Algorithm

LM is used for solving nonlinear, least-squared problems. Like the other fitting algorithm, LM finds only a local minimum rather than a global minimum. It is interpolated with the gradient descent and Gauss-Newton (GN) algorithm. LM is more efficient than GN. The primary application of LM is in the squares curve-fitting problem.

Steps involved during the training of the MI dataset:

Input $f: R^n \to R$ a function such that

$$f(x) = \sum_{i=1}^{m} (f_i(x))^2 \tag{10.8}$$

In Eq. (10.8) f_i are differentiable functions from R^n to R, x^0 is an initial solution, Output x^*, and a local minimum of the cost function f

Step 1: Initialize the k value which is 0
Step 2: Compute max diag $(J^T J)$ and pass the value to λ
Step 3: $x = x^0$
Step 4: If the STOP-CRIT and $(k < k_{max})$
    Compute, find δ such that $J^T J + \lambda\mathrm{diag}(J^T J))\delta = J^T J$
    $x' = x + \delta$ if the value of $f(x') < f(x)$ then $\lambda = \frac{\lambda}{r}$
    If the value of $f(x') > f(x)$ then
    $\lambda = v\lambda$ and $k = k + 1$

The basic idea of LM is to perform a combined training process around areas with complex curvatures.

10.3.5.1   Result and Discussion

The research work demonstrated used the sample size of 1024 data that includes MI patient's short axis delayed enhancement region of the heart MRI and the normal heart MRI. The dataset has been spilt into 716 data for training, 154 data for validation, and 154 for testing. The data provides a platform for the particular classification of the attribute. The data which were used in this research work is from the Sunny Brook dataset. The MICCAI workshop challenge in 2009 used the subset of this data for

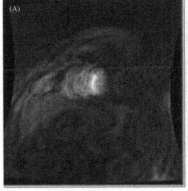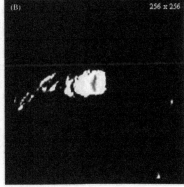

FIGURE 10.4   (A) Original grayscale image of the MI. (B) Segmented image using the FCM (multispectral and single channel).

automatic myocardial segmentation in the short axis using MRI. A lot of work with this dataset happened prior to its usage. The implementation of the classification in the research work yields various steps.

10.3.5.2 Phase I

It is often necessary to do segmentation for the identification of scar tissue in short axis delayed enhancement MRI. Segmentation is the important aspect of medical image processing to identify defected regions. The FCM algorithm has proved to be the superior among the other clustering techniques in terms of efficiency. FCM reaches efficiency by improving the cluster center and changes in the membership value. In this research work, the FCM was used as the segmentation technique to enhance efficiency.

Fig. 10.4A shows the original grayscale image and Fig. 10.4B shows the segmentation image result by the FCM multispectral and the single channel. Finally, after successful running, the FCM using the multispectral segmentation algorithm, it may be helpful to twist some parameters and re-run the algorithm with single channel for better identification. FCM is considered an efficient algorithm for the image segmentation. However, it has the disadvantages of insufficient robustness to image noise and sensitivity to outlier, that is, the Euclidean distance in FCM. So, we used the FCM single channel to further reduce the noise in the image.

10.3.5.3 Phase II

The segmentation result is then passed through the morphological filtering technique for feature extraction. Fig. 10.5A shows the output image of the FCM segmentation and Fig. 10.5B shows the morphological filtering technique for the morphology of the features in the segmented image.

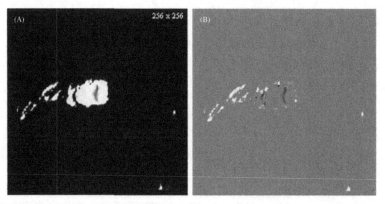

FIGURE 10.5 (A) Output of the FCM segmentation. (B) Morphological image filtering technique for ROI.

10.3.5.4   Phase III

The extracted features are trained using the LMBP in the FFNN for the classification of the MI delayed enhancement MRI. The MI data is loaded into the array inputs and the MI data output targets into the target arrays. The network used for training the FFNN is LMBP along with the sigmoid transfer function. Ten neurons are assigned for one hidden layer. The research work uses 10 hidden layers. The output has one neuron because the target value assigned is 1. More neurons are required for the complex problems. More layers require more computation. The division of the dataset in the research work is 70% for the training data, 15% for the validation data, and 15% for the testing data. So, for the training, 176 data have been allocated in the research work, 154 data for validation, and 154 for testing.

The training for the network is done using the LMBP and SCG. The performance is calculated using the mean-squared-error (MSE). The training stops when the validation error increases for six iterations, which occurs at the iteration point of 20, as shown in Fig. 10.6. Fig. 10.6A shows the regression, Fig. 10.6B represents the error histogram, where the blue level indicates the training data, the green level indicates the validation data, and the red level indicates the testing data. The histogram gives an indication of the outlier, which are the data points and specifies the worst, rather than the majority, of the data. These outliers are also visible in the testing regression plot.

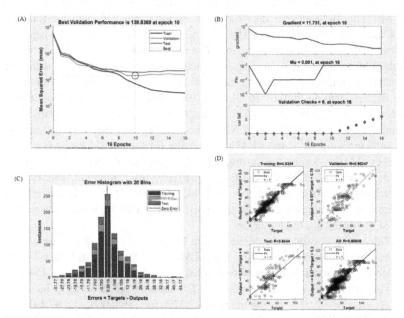

FIGURE 10.6   (A) Regression in the training window. (B) Gradient descent. (C) Mean Squared Error window. (D) Training error using a histogram.

TABLE 10.1 Results of Levenberg-Marquard Back Propagation Algorithm and Scaled Conjugate Gradient Algorithm With Six Hidden Layers

Training Data			
Algorithm	Samples	MSE	Regression
LMBP algorithm	716	128.62	93.53
Scaled conjugate gradient	716	128.62	86.33
Validation Data			
LMBP algorithm	154	239.52	86.24
Scaled conjugate gradient	154	239.52	81.10
Testing Data			
LMBP algorithm	154	237.34	84.43
Scaled conjugate gradient	154	237.34	79.38

Train the data repeatedly using LMBP for an increase in accuracy. The two-layered FFNN is the standard network used for functional fitting. One is the sigmoid transfer function in the hidden layers and the other is the linear transfer function in the output layer. The research included 10 hidden layers.

Table 10.1 shows the results of the LMBP and SCG for the classification of MI. The 1024 sample data has been divided into 716 data for training, 154 data for validation, and 154 for the testing data. The performance is tested with the MSE and regression. The performance result is shown in the table. The values of the measures in function of training the FFNN using the input images that present to the network and the target image data defining the desired network output. Randomly, 1024 data samples were divided into 716 data for training, 154 data for validation, and 154 data for testing. The training based on LMBP for FFNN yields 93.53% accuracy for the training, 86.24% for validation, and 84.43% for testing for the six hidden layers. MSE reaches 237.34 during the testing. The training based on SCG for FFNN yields 86.33% accuracy for the training, 81.10% for validation, and 79.38% for testing the six hidden layers. The MSE reaches 237.34 during testing. These comparative results show that the LMBP yields higher accuracy than the SCG for the six hidden layers.

Table 10.2 illustrates the Mean Square Error (MSE) and Regression values for training, validation and testing using LMBP and SCG for 10 hidden layers. Randomly, 1024 data samples were divided into 716 data for training, 154 data for validation, and 154 data for testing. The training based on SCG increases the accuracy to 97.28% for the training, 97.32% for validation, and 97.14% for testing. The MSE reaches 33.76 during the testing. The trained network with the 10 hidden layers based on LMBP increases the accuracy to 99.36% for the

TABLE 10.2 Results of Levenberg-Marquard Back Propagation Algorithm and Scaled Conjugate Gradient Algorithm With 10 Hidden Layers

Training Data			
Algorithm	Samples	MSE	Regression
Levenberg-Marquard Back Propagation algorithm	716	1.45	99.36
Scaled conjugate gradient	716	29.28	97.28
Validation Data			
Levenberg-Marquard Back Propagation algorithm	154	1.86	98.80
Scaled conjugate gradient	154	31.96	97.32
Testing Data			
Levenberg-Marquard Back Propagation algorithm	154	1.44	99.01
Scaled conjugate gradient	154	33.76	97.14

TABLE 10.3 Results of Levenberg-Marquard Back Propagation Algorithm and Scaled Conjugate Gradient Algorithm With Retrained 10 Hidden Layers

Training Data			
Algorithm	Samples	MSE	Regression
Levenberg-Marquard Back Propagation algorithm	716	1.45	99.98
Scaled conjugate gradient	716	11.89	98.96
Validation Data			
Levenberg-Marquard Back Propagation algorithm	154	1.86	99.98
Scaled conjugate gradient	154	15.08	98.63
Testing Data			
Levenberg-Marquard Back Propagation algorithm	154	1.44	99.10
Scaled conjugate gradient	154	18.38	98.30

training, 98.80% for validation, and 99.01% for testing. The MSE reaches 99.93 during the testing. The above comparative results reveal that the LMBP achieves higher accuracy than the SCG for the 10 hidden layers.

Table 10.3 illustrates the values for repeated training through the FFNN using the input images that present to the network and the target image data defining the desired network output. Randomly, 1024 data samples were divided into 716 data for training, 154 data for validation, and 154 data for testing. The retrained network is based on SCG which increases the accuracy to 98.9% for the training, 98.6% for validation, and 98.3% for testing. The MSE reaches 18.38 during testing. SCG requires less memory. Training automatically stops when generalization stops improving, as indicated by an increase in the MSE of the validation samples. The trained network with the 10 hidden layers are repeatedly trained using LMBP on FFNN which increases the accuracy to 99.99% for the training, 99.98% for validation, and 99.98% for testing. The MSE reaches 99.10 during testing. This repeated training on the LMBP and SCG for FFNN results shows that the LMBP increases the accuracy on repeated training using 10 hidden layers.

Fig. 10.6A shows the MSE performance at the 16 epoch. An epoch is the measure of the number of times the training vectors are used. Once for updating the weights. Fig. 10.6B shows the gradient descent in context of learning the back propagation for the adjustment of weight of the neurons by calculating the loss function of the gradient. The gradient value is 11.73 at

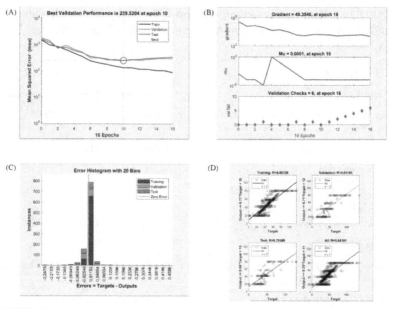

FIGURE 10.7 (A) Regression in the training window. (B) Gradient descent. (C) MSE window. (D) Training error using a histogram.

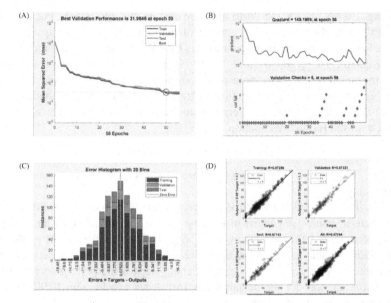

FIGURE 10.8   (A) Regression in the training window. (B) Gradient descent. (C) MSE window. (D) Training error using a histogram.

the epoch 16. The control parameter for the algorithm is mu which is equal to 0.001. To prevent the training network from performing poorly on nontraining, the training stops if the performance on the validation degrades for the default value, that is, six epochs. The procedure is called the validation stop. Fig. 10.6C shows the error histogram with 20 bins and is calculated by:

$$Errors = Target - Outputs$$

The default value of the error histogram is 20 bins. Fig. 10.6D shows the regression percentage for the training, validation, and testing.

Fig. 10.7A shows the MSE performance at the 16 epoch. Fig. 10.7B shows the gradient descent in context of learning the back propagation for the adjustment of weight of the neurons by calculating the loss function of the gradient. The gradient value is 48.35 at epoch 16. The control parameter for the algorithm is mu which is equal to 0.0001. To prevent the training network on performing poorly on nontraining, the training stops if the performance on the validation degrades for the default value, that is, six epochs. The procedure is called the validation stop. Fig. 10.7C shows the error histogram with 20 bins and it is calculated by:

$$Errors = Target - Outputs$$

The default value of the error histogram is 20 bins. Fig. 10.7D shows the regression percentage for the training, validation, and testing.

Fig. 10.8A shows the MSE performance at the 56 epoch. Fig. 10.8B shows the gradient descent in context of learning the back propagation for the adjustment of weight of the neurons by calculating the loss function of the gradient. The gradient value is 149.198 at the epoch 56. To prevent the training network on performing poorly on nontraining, the training stops if the performance on the validation degrades for default value, that is, six epochs. Fig. 10.8C shows the error histogram with 20 bins and it is calculated with the following equation.

$$\text{Errors} = \text{Target} - \text{Outputs}$$

Fig. 10.8D shows the regression percentage for the training, validation, and testing.

Fig. 10.9A shows the MSE performance at the 62 epoch. Fig. 10.9B shows the gradient descent in context of learning the back propagation for the adjustment of weight of the neurons by calculating the loss function of the gradient. The gradient value is 103.012 at epoch 62. To prevent the training network on performing poorly on nontraining, the training stops if the performance on the validation degrades for the default value, that is, six epochs. Fig. 10.9C shows the error histogram with 20 bins and it is calculated by:

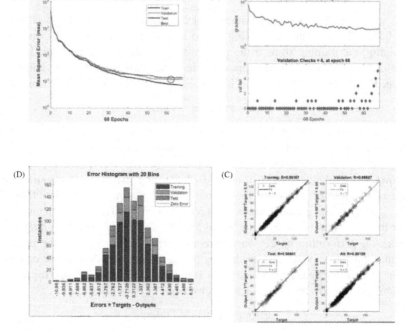

FIGURE 10.9 (A) Regression in the training window. (B) Gradient descent. (C) MSE window. (D) Training error using histogram.

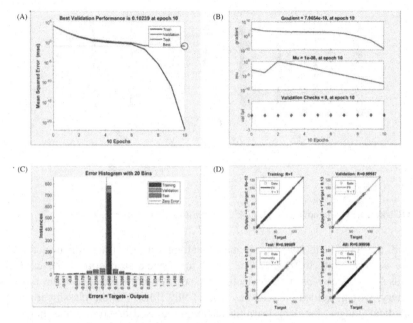

FIGURE 10.10   (A) Regression in the training window, (B) gradient descent, (C) MSE window, (D) training error using histogram.

$$Errors = Target - Outputs$$

Fig. 10.9D shows the regression percentage for the training, validation, and testing.

Fig. 10.10A shows the MSE performance at the 10 epoch. Fig. 10.10B shows the gradient descent in context of learning the back propagation for the adjustment of weight of the neurons by calculating the loss function of the gradient. The gradient value is 7.96 at the epoch 10. To prevent the training network on performing poorly on nontraining, the training stops if the performance on the validation degrades for the default value, that is, six epochs. Fig. 10.10C shows the error histogram with 20 bins and it is calculated by:

$$Errors = Target - Outputs$$

Fig. 10.10D shows the regression percentage for the training, validation, and testing.

Fig. 10.11 shows the existing classification performance and is compared with the proposed work. The compared result explicitly shows that the proposed framework yields high accuracy for the MI dataset. The probabilistic neural network with laws' mask analysis along with filter [40] which produces the accuracy rate of 72.5% and the process has been tested with the mammographic breast images. In the same way, the process has been tested

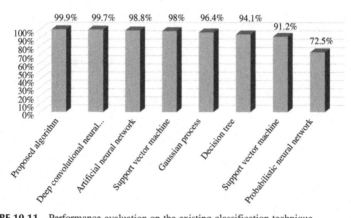

FIGURE 10.11   Performance evaluation on the existing classification technique.

with the SVM with laws' mask analysis along with the filter which increases the accuracy rate to 91.2%. The SVM for the classification of the cardiac muscle, muscular artery, connective tissue, the larger vein, and the elastic artery, which yielded an accuracy rate of 98% [41]. For the classification of medical images, the deep-learning technique is used based on datasets of 24 classes for the five modalities and yields a classification accuracy of 99.7% [42]. The ANN is proposed [43] for the classification of real-time cholesterol with biomedical images achieving an accuracy rate of 98.8%. The proposed diagnostic decision tree [44] for distinguishing Cerebellar ataxias grounded on the MRI yields accuracy of 94.1%. The Gaussian process (GP) has been proposed for the classification of super paramagnetic relaxometry data. The GP result is compared with the reconstruction technique and the result reveals that the GP outperforms with an accuracy rate of 96.4% [45]. Above all, the proposed model of this chapter is based on the FCM for the segmentation, morphological filtering technique for the feature extraction, and, finally, using the FFNN technique trained using the LMBP and yields a high accuracy rate of 99.9%. Therefore, the proposed method is considered as one of the best methods for the classification of the MI short axis delayed enhancement MRI.

10.4   CONCLUSION

The classification of myocardial ischemia with short axis delayed enhancement in MRI using a machine-learning technique yields better accuracy than other classification algorithms. The proposed model uses FCM for the segmentation to identify the scar region in the myocardial ischemia patient's delayed enhancement MRI and the feature extraction is done with the help of the morphological process. Finally, the model is trained using the Levenberg−Marquardt back propagation in FFNN which is used to classify the myocardial ischemia. The proposed model is compared with the other classification algorithms and the results reveal that the proposed model produces higher accuracy on the classification of myocardial ischemia. Therefore, the proposed model is one of the most viable techniques for classifying the MI in short axis delayed enhancement MRI.

REFERENCES

[1] H.J. Arevalo, F. Vadakkumpadan, E. Guallar, A. Jebb, P. Malamas, K.C. Wu, et al., Arrhythmia risk stratification of patients after myocardial infarction using personalized heart models, Nat. Commun. (7) (2016).

[2] C. Ramanathan, N.G. Raja, P. Jia, K. Ryu, Y. Rudy, Noninvasive electrocardiographic imaging for cardiac electrophysiology and arrhythmia, Nat. Med. 10 (4) (2004).

[3] P. Rajiah, M.Y. Desai, D. Kwon, S.D. Flamm, MR imaging of myocardial infarction, Radiographics 33 (5) (2013) 1383–1412.

[4] W. Zhao, X. Xu, Y. Zhu, F. Xu, Active contour model based on local and global Gaussian fitting energy for medical image segmentation, Optik 158 (2018) 1160–1169.

[5] G. Nagarajan, R.I. Minu, B. Muthukumar, V. Vedanarayanan, S.D. Sundarsingh, Hybrid genetic algorithm for medical image feature extraction and selection, Proced. Comput. Sci. 85 (2016) 455–462.

[6] G. Mohan, M. Monica Subashini, MRI based medical image analysis: survey on brain tumor grade classification, Biomed. Signal Process. Control 39 (2018) 139–161.

[7] A.W. Simonetti, W.J. Melssen, M. van der Graaf, G.J. Postma, A. Heerschap, L.M.C. Buydens, A chemometric approach for brain tumor classification using magnetic resonance imaging and spectroscopy, Anal. Chem. 75 (2003) 5352–5361.

[8] T. Tantimongcolwat, T. Naenna, C. Isarankura-Na-Ayudhya, M.J. Embrechts, V. Prachayasittikul, Identification of ischemic heart disease via machine learning analysis on magnetocardiograms, Comput. Biol. Med. 38 (7) (2008) 817–825.

[9] S.R. Kannan, S. Ramathilagam, R. Devi, A. Sathya, Robust kernel FCM in segmentation of breast medical images, Expert Syst. Appl. 38 (4) (2011) 4382–4389.

[10] A.P. Nanthagopal, R.S. Rajamony, Classification of benign and malignant brain tumor CT images using wavelet texture parameters and neural network classifier, Int. J. Comput. Vis. (2012). Springer.

[11] L. He, Y. Wen, M. Wan, S. Liu, Multi-channel features based automated segmentation of diffusion tensor imaging using an improved FCM with spatial constraints, Neurocomputing 137 (2014) 107–114.

[12] S.K. Adhikari, J.K. Sing, D.K. Basu, M. Nasipuri, Conditional spatial fuzzy c-means clustering algorithm with application in MRI image segmentation, Inf. Syst. Des. Intell. Appl., Springer, 2015, pp. 539–547.

[13] X. Gao, S. Lin, T.Y. Wong, Automatic feature learning to grade nuclear cataracts based on deep learning, IEEE Trans. Biomed. Eng. 62 (1) (2015) 2693–2701.

[14] A. Payan, G. Montana, arXiv:1502.02506 Predicting Alzheimer's Disease: A Neuroimaging Study with 3D Convolutional Neural Networks, arXiv preprint, 2015.

[15] W. Shen, M. Zhou, F. Yang, C. Yang, J. Tian, Multi-scale convolutional neural networks for lung nodule classification, International Conference on Information Processing in Medical Imaging, Springer, 2015, pp. 588–599.

[16] J. Antony, K. McGuinness, N.E. O'Connor, K. Moran, Quantifying radiographic knee osteoarthritis severity using deep convolutional neural networks, 2016 23rd International Conference on Pattern Recognition (ICPR), IEEE, 2016, pp. 1195–1200.

[17] E. Hosseini-Asl, G. Gimel'farb, A. El-Baz, arXiv:1607.00556 Alzheimer's Disease Diagnostics by a Deeply Supervised Adaptable 3D Convolutional Network, arXiv preprint, 2016.

[18] J. Kawahara, G. Hamarneh, Multi-resolution-tract CNN with hybrid pretrained and skin-lesion trained layers, International Workshop on Machine Learning in Medical Imaging, Springer, 2016, pp. 164–171.

[19] M.S. Choudhry, R. Kapoor, Performance analysis of fuzzy C-means clustering methods for MRI image segmentation, Proced. Comput. Sci. 89 (2016) 749–758.

[20] J. Cao, L. Chen, M. Wang, H. Shi, Y. Tian, A parallel adaboost-backpropagation neural network for massive image dataset classification, Sci. Rep. (6) (2016).

[21] A. Danielis, D. Giorgi, M. Larsson, T. Strömberg, S. Colantonio, O. Salvetti, Lip segmentation based on Lambertian shadings and morphological operators for hyper-spectral images, Pattern Recognit. 63 (2017) 355–370.

[22] R.S.A. Gupta Deep, A hybrid edge-based segmentation approach for ultrasound medical images, Biomed. Signal Process. Control 31 (2017) 116–126.

[23] S.-C. Yang, A robust approach for subject segmentation of medical images: illustration with mammograms and breast magnetic resonance images, Comput. Electr. Eng. 62 (2017) 151–165.

[24] R.R. Gharieb, G. Gendy, A. Abdelfattah, H. Selim, Adaptive local data and membership based KL divergence incorporating C-means algorithm for fuzzy image segmentation, Appl. Soft Comput. 59 (2017) 143–152.

[25] H.-X. Pei, Z.-R. Zheng, C. Wang, C.-N. Li, Y.-H. Shao, D-FCM: density based fuzzy c-means clustering algorithm with application in medical image segmentation, Procedia Comput. Sci. 122 (2017) 407–414.

[26] D. Ashok Kumar, V. Kumar, R. Kumar, Detection of myocardial infarction in 12 lead ECG using support vector machine, Appl. Soft Comput. 64 (2018) 138–147.

[27] L. Guo, P. Wang, R. Sun, C. Yang, N. Zhang, Y. Guo, et al., A fuzzy feature fusion method for auto-segmentation of gliomas with multi-modality diffusion and perfusion magnetic resonance images in radiotherapy, Sci. Rep. 8 (1) (2018).

[28] J. Li, W. Tang, J. Wang, X. Zhang, Multilevel thresholding selection based on variational mode decomposition for image segmentation, Sig. Process. 147 (2018) 80–91.

[29] B. Devkota, A. Alsadoon, P.W.C. Prasad, A.K. Singh, A. Elchouemi, Image segmentation for early stage brain tumor detection using mathematical morphological reconstruction, Proced. Comput. Sci. 125 (2018) 115–123.

[30] F. Zhao, J. Zhao, W. Zhao, F. Qu, L. Sui, Local region statistics combining multi-parameter intensity fitting module for medical image segmentation with intensity inhomogeneity and complex composition, Opt. Laser Technol. 82 (2016) 17–27.

[31] Z.-X. Ji, Q.-S. Sun, D.-S. Xia, A framework with modified fast FCM for brain MR images segmentation, Pattern Recognit. 44 (5) (2011) 999–1013.

[32] Z.M. Wang, Y.C. Soh, Q. Song, K. Sim, Adaptive spatial information-theoretic clustering for image segmentation, Pattern Recognit. 42 (9) (2009) 2029–2044.

[33] J. Hu, T. Li, C. Luo, H. Fujita, Y. Yang, Incremental fuzzy cluster ensemble learning based on rough set theory, Knowl.-Based Syst. 132 (2017) 144–155.

[34] S. Mitra, W. Pedrycz, B. Barman, Shadowed c-means: integrating fuzzy and rough clustering, Pattern Recognit. 43 (4) (2010) 1282–1291.

[35] T. Chaira, A novel intuitionistic fuzzy C means clustering algorithm and its application to medical images, Appl. Soft Comput. J. 11 (2) (2011) 1711−1717.

[36] Y.K. Dubey, M.M. Mushrif, Segmentation of brain MR images using intuitionistic fuzzy clustering algorithm, Proceedings of the 8th Indian Conference on Computer Vision, Graphics and Image Processing (ICVGIP '12), ACM, New York, 2012.

[37] Z. Yamin, Z. Mengmeng, G. Xiaomin, Z. Zhiwei, Z. Jianhua, Research on matching method for case retrieval process in CBR based on FCM, Proced. Eng. 174 (2017) 267−274.

[38] J.P. Thiran, B. Marq, Morphological feature extraction for the classification of digital images of cancerous tissues, IEEE Trans. Biomed. Eng. 43 (10) (1996) 1011−1020.

[39] W. Cao, X. Wang, Z. Ming, J. Gao, A review on neural networks with random weights, Neurocomputing 275 (2018) 278−287.

[40] I. Kumar, J. Virmani, H.S. Bhadauria, M.K. Panda, Kriti, Classification of breast density patterns using PNN, NFC, and SVM classifiers, Soft Comput. Based Med. Image Anal. (2018) 223−243.

[41] C. Mazo, E. Alegre, M. Trujillo, Classification of cardiovascular tissues using LBP based descriptors and a cascade SVM, Comput. Methods Progr. Biomed. 147 (2017) 1−10.

[42] A. Qayyum, S.M. Anwar, M. Awais, M. Majid, Medical image retrieval using deep convolutional neural network, Neurocomputing 266 (2017) 8−20.

[43] K.G. Adi, P.V. Rao, V.K. Adi, Analysis and detection of cholesterol by wavelets based and ANN classification, Proced. Mater. Sci. 10 (2015) 409−418.

[44] M. Higashi, K. Ozaki, T. Hattori, K.S. Takashilshii, N. Sato, M. Tomita, et al., A diagnostic decision tree for adult cerebellar ataxia based on pontine magnetic resonance imaging, J. Neurol. Sci. 387 (2018) 187−195.

[45] J. Sovizi, K.B. Mathieu, S.L. Thrower, W. Stefan, J.D. Hazle, D. Fuentes, Gaussian process classification of superparamagnetic relaxometry data: phantom study, Artif. Intell. Med. 82 (2017) 47−59.

Chapter 11

Simple-Link Sensor Network-Based Remote Monitoring of Multiple Patients

S. Vishnu[1], S.R. Jino Ramson[1], K. Lova Raju[1] and Theodoros Anagnostopoulos[2]

[1]Department of Electronics and Communication Engineering, Vignan's Foundation for Science, Technology and Research, Guntur, India, [2]Department of Infocommunication Technologies, ITMO University, St. Petersburg, Russia

Chapter Outline

11.1 INTRODUCTION

Chronic diseases are noncommunicable diseases (NCD) that cannot be cured, but can be controlled. As lifestyles improve, the probability of being affected by chronic diseases is high. The world statistics of the people affected by chronic diseases is shown in Fig. 11.1; it is observed that the largest proportion of NCD death is by cardiovascular disease (48%) followed by cancer (21%), chronic respiratory conditions (12%), and 19% by other chronic diseases. If no active care is taken, there is a great chance that the affected person may suffer from severe pain and, sometimes, death.

Recent research and development in the wireless network field have led to a new and innovative approach in the medical field. Continuous

Intelligent Data Analysis for Biomedical Applications. DOI: https://doi.org/10.1016/B978-0-12-815553-0.00012-4

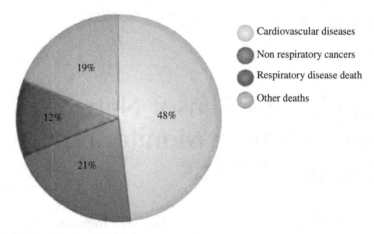

FIGURE 11.1 Statistics of death due to chronic diseases.

health monitoring is required in the case of many chronic diseases and regular checking of physiological data is essential in these cases. But it need not be required to be hospitalized for the sake of continuous monitoring purposes. Moreover, it is expensive to stay in the hospital and consumes the patients' valuable time. This situation can be addressed by a wireless sensor networks-based health monitoring system [1,2], consisting of sensors, a transmitter, receiver, and a personal computer (PC) to monitor the physiological signals. Sensors are used to detect the physiological parameters, using tools such as electrocardiography (ECG), blood pressure tests, etc. This data can be transmitted to a central node (probably a PC) located in hospitals, which can be analyzed by a clinical professional or an artificial intelligence algorithm developed for the diagnosis. Real-time monitoring and decision-making allows the patient to get the right care at the right time. Much advancement is happening in the field of remote health monitoring systems (RHMS). Initially, it was just a messaging system based on global system for mobile (GSM) or general packet radio services (GPRS). But the technological advancements in the field of wireless communication allows for highly accurate diagnoses of patients' health conditions. Investment in remote monitoring technologies has made an incredible change in the field of medical science. This led to the reduction in long stays and lengthy queues in hospitals and it had also made it easier for patients to reach clinicians in a very short time. The system described in this chapter is very helpful for people suffering from chronic diseases because it connects the patient with the hospital in real-time. This chapter describes the development of a remote monitoring system for ECG.

ECG is the process of recording the heart rate and its variations by using an external electrode that is attached to the skin. The two electrodes present in the device are used to obtain the voltage difference recorded by the equipment. ECG signals generally consist of a low amplitude signal in the presence of high offset voltage and noise. The main sources of noise in ECG signals are muscle noise and power line radiation, etc. Since the ECG signal has a low amplitude compared to the noise, it is difficult to distinguish both signals. To overcome this, an analog frontend processor having an instrumental amplifier and a noise-cancellation circuit is provided. After filtering, the ECG signals are fed into the microcontroller for further processing. An ECG signal mainly consists of three types of waves: P wave, QRS wave, and T wave. A basic ECG signal is shown in Fig. 11.2.

The P wave represents the atrial depolarization. The duration of this wave is less than 0.12 s and its amplitude range is less than 2.5 mV. A P wave with increased amplitude indicates the enlargement of the right atrium and a decreased amplitude indicates hyperkalemia, and a QRS wave represents the depolarization of the right and left ventricles. Normally QRS wave lasts for 0.06−0.10 s. The amplitude of this wave is 3.5 mV. A T wave represents the repolarization of ventricles. A negative T wave indicates coronary ischemia. As this indicates the repolarization of ventricles, the wave contains more information compared to a P wave and QRS wave.

This chapter proposes a wireless sensor-based remote monitoring system to monitor the ECG signals of a patient. Each patient will act as a node and the hospital section is the central node. The patient's ECG signal can be sent to the central node with the help of a wireless access point provided in the home. Once the signals are processed in the hospital section the clinician can make an appropriate decision at the right time. The elimination of a PC to send data to the hospital section and the usage of CC3200 (a single chip wireless microcontroller with Internet protocol (IP) network module) to reduce intermediate nodes in the home section to establish the connection are the highlights of this research. The current consumption, life expectancy, and maximum range of the home unit from the wireless access point are also evaluated for the proposed system. Many types of research are performed in

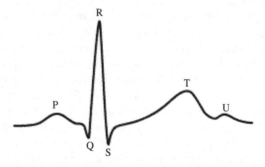

FIGURE 11.2   Basic electrocardiography (ECG) signal.

the biomedical field using various technologies, such as PC, Bluetooth, and smartphones, etc. The real-time monitoring of patients has many advantages, such as helpings to reduce the chance of critical conditions and sudden variations in the physiological signals.

The rest of the chapter is organized as follows, Section 11.2 explains the related works, Section 11.3 explains the scenario of the proposed system by describing the patient's home unit and the hospital section, Section 11.4 describes the results and discussions along with the calculations of power consumption and life expectancy of the WPSU. Finally, conclusions are drawn in Section 11.5.

11.2 RELATED WORKS

The study of various existing monitoring systems is discussed in this section. Tagad et al. [3] proposed a mobile, wearable ECG detector using Bluetooth technology to establish a connection from the detector. The system can be used at home and is useful in emergency situations to monitor patients and to save time. Its architecture has three sections: ECG detector, PC/Palmtop computer, and a server at a hospital. The ECG detector is equipped with signal sampling preprocessing, amplifying capabilities, and AD transferring modules. Bluetooth is used to establish the communication between the ECG detector and the Palmtop/PC. The detector accepts low-voltage ECG signals and these signals are sent to the frontend amplifier which amplifies the low-voltage signal. The amplified signal is then converted into digital packets and are sent to the microcontroller. Microchip PIC18F452 is used as the controller. All the data are processed through the Bluetooth module and sent to the Palmtop/PC. Bluetooth interface in Palmtops work with L2CAP and the detector works with Host Controller Interface. The information received in the Palmtop/PC through Bluetooth is further transmitted to the server at the hospital by using GPRS. The main disadvantage of using this system is that Bluetooth has a limited range. As the range of Bluetooth is increased, the cost is increased. Another drawback is the use of GPRS. Once the call is connected it is not possible to send information using GPRS.

A ZigBee protocol-based system for remote monitoring is proposed in Refs. [4,5]. A wireless connection is achieved within a range of 50 m. Cai Ken et al. [6] present a remote medical monitoring system for ECG data. The system consists of two parts, the portable remote medical monitoring unit and the monitoring centre. A reduced instruction set computer (RISC)-based arm processor, embedded OS custom ECG detector module, and GPRS constitute the remote monitoring unit. The monitoring station can perform real-time analysis of a patient's data wirelessly with the help of an information processing system. The main objective of this system is to provide a warning mechanism to the patient using the analyzed data and to provide medical assistance irrespective of time and place. This work concentrates only on ECG signals, but it can be implemented to measure any physiological parameters. In some systems, a graphical user interface is made to show the physiological parameters [7]. The parameters considered here are ECG, heart rate, SPO2, and temperature. After processing the signals, it will be available in graphical user interface (GUI). In this research, GSM technology is used for the transmission of the patient's data to the server for monitoring purposes. So, even in remote sites expert healthcare can be given if internet facility is available. Also, provision is made to send an SMS to the doctor if the concerned patient's temperature, heartbeat, or HbO_2 crosses the limited range. This system using a GSM code to obtain the features of the signals, which is more costly. Limitations include GSM's limited data rate capability.

Sukanesh et al. [8] proposed a system to predict cardio diseases in the early stage. The system has a database for comparing the ECG signals of the patient continuously. If any variations from the threshold limit are raised, then immediate information will be sent to the doctor. The highlight of this work is the alarm system implemented for the immediate help of the doctor if any abnormalities occur. The use of PC or laptop in the biosignal acquisition module is a drawback for this system.

Noureddine et al. [9] presented a system in which the monitoring of ECG is achieved by using a portable wireless device. The real-time electrical activity of the heart can be sent to a mobile phone [10,11] or a PC with the help of the wireless transmission capabilities of the device. This low-cost system can perform acquisition, processing, storing, and visualization of the collected information. The system consists of ASIC which has integrated analog-to-digital converter (ADC) and serial communication ports. A low power multisensory frontend acquisition system is used for capturing the ECG signal.

The main challenges in the monitoring of noninvasive long-term breathing problems are analyzed in Ref. [12]. A miniaturized breathing detector is developed to detect breathing and its cessation. It also provides an automated algorithm to analyze the breathing signal. As the detector is small, the signal quality is affected as it contains noise signals along with the actual signal. It works by using a battery which is size and light so that the patient can wear it. By using the miniaturized detector, the oximetry and the impedance across the chest are calculated by providing the required input signal. The detector detects the breathing signal which also contains noise signals. Acoustic sensing is chosen due to the presence of many noise signals. The size, shape, and weight of the sensor are defined in such a way that the patient can carry it on their body. The sensor is mounted on the neck so that a signal with large signal power is obtained. The main problem faced by this design is the complexity of the algorithm used, and that it increases power consumption. Due to high-power consumption, the lifetime of the battery is very low.

One main area under consideration for portable or remote application is power consumption. Ng and Chan et al. [13] proposed a CMOS-based analog frontend which is programmable for ECG and electoencephalography (EEG) signal measurement. Without sacrificing the signal quality, the proposed programmable CMOS AFE IC can perform the task with a reduced supply voltage. As a result, even button cell batteries can be used as the power supply. The use of AFE IC results in low power consumption and compact design. The highlights of AFC IC are high CMRR and reduction in input referred noise, etc.

Heart beats with an irregular rhythm is called arrhythmia [14]. To detect arrhythmia, a wearable ECG sensor is proposed in Ref. [15]. In this work, an ECG sensor is associated with a continuous event recorder. The two electrodes of the sensor are directly placed in the patient's skin. The low voltage ECG signal is then amplified with the help of electrical circuit connected to the sensor. The amplified signal will be transmitted to a handheld device (HHD). The receiver is capable of analyzing and storing all the details received from the sensor circuit. If any variations are found from threshold values, then the received data will be sent to the hospital's server with the help of GPRS. At the hospital side, diagnosis will be performed with help of a WPR client installed PC. The sampled ECG signals are fed to a personal digital assistant (PDA), which has a CF-slot GSM/GPRS module. The software in the PDA allows the automatic connection to GPRS. The Internet server will be updated with the data received via GPRS. The doctor at the hospital can retrieve all the information and diagnose the problems. The advantages of using this system include that it is easy to use, the sensor can be replaced by patients themselves, and it has an inbuilt alarm detector. The drawback in using this system is the cost required for the handheld devices (HDD) and the limitations of GPRS.

Focusing on providing an emergency alert by analyzing the recorded patient's data, a mobile phone-based system was proposed in [16]. Since a mobile phone is used in the system, separating GSM or GPS modules is not required. Peer-to-peer communication is possible with the help of Bluetooth android API. The sensor is with a Bluetooth module and the collected data will be sent to the android phone via Bluetooth [17]. The Android phone will update the received data to the server. The patient's data will be available on the server and the physician will be able to diagnose the patient's situation. Deshmukh et al. [18] proposed a system which can monitor a patient's heart beat and body temperature. Remote monitoring is achieved by Wi-Fi communication. A website is created for the real-time monitoring of medical parameters. Open standard transportation protocol is used to give warning and notification to the doctor regarding the patient's health condition. The data can also be sent to the doctor's laptop/PC as email by WSP. The monitoring and suggestions of doctors worldwide can be obtained in this system.

Shebi et al. [19] proposed a smartphone-based health monitoring system. ECG signals are obtained with two lead ECG sensors. The data obtained is transmitted via a wireless link. A smartphone is used for processing and to provide GUI (to provide ECG waveforms). This system has three units: a concerto microcontroller with an ECG acquisition module, simple link CC3000 module unit, and an application platform for a smartphone. Due to the onboard powerful computing capability, the smartphone is used as a handheld computer. The highlight of this system is its simplicity in use and its affordability. It not only reduces power consumption, but also avoids the complexity of a hard-wired link. The main challenge discussed in this chapter is to develop a portable device to analyze and store physiological signals and to transmit these signals over a wireless connection [20]. The developed system has a graphic online display and is capable of digital transmission of the full signals. Most of the devices discussed currently lack an online display. Since the developed signal has an online display it is easy to integrate the signal to a PDA. The signal transmitted to the PDA is stored and analyzed by experts at the hospital. The details stored can be used later when an offline analysis is required. The entire system is implemented by using a commercially available HP Pocket PC with NI Lab-VIEW Mobile Module installed [21], CF-6004, and a frontend ECG signal conditioner. The signal conditioning circuit prevents the patient from being exposed to electrical currents. The conditioned signal is then fed to CF-6004 which scans the signal and gives an accurate measurement. LabVIEW mobile module is used to write the entire PDA application. It consists of three tasks: the first is data acquisition and storing the data, the second task is to establish a wireless connection between the base station and the laptop/PC, and the third task is to store all the received data in PDAs' memory locally. Since the display of PDA is very small, a tabbed user interface display has been developed.

Dilmaghani et al. [22] presented a system which can measure the physiological signals from multiple patients from their homes. In this system, the need for a PC in the home is removed [23], which is an improvement compared with other systems which require a PC in the home. Each patient is considered as a node and a home unit will be installed in the patient's home for monitoring physiological parameters. The system mainly concentrates on patients suffering from chronic diseases. All the nodes (patients) are connected to a central node which is installed at the hospital. The system consists of two units: one is home unit and another one is hospital unit. The home unit has an ECG sensor, wireless patient portable unit (WPPU), and wireless access point unit (WAPU). WPPU is placed on the patient's body and the ECG sensor will provide the signal to WPPU by using a very light set of wires. Once signals are obtained, WPPU and WAPU will provide these signals to the central node. The main disadvantage of this system is the intermediate node between WPPU and the home router. Also, the connection between WAPU and the home router is wired.

By considering all the drawbacks obtained from the existing systems discussed here, this chapter proposes a new system which uses wireless components for measuring signals. In the proposed method, the patient's home requirements consist of only Wireless Portable Sensor Unit (WPSU) and a home router, so the main objectives of the proposed system is to use a single-chip microcontroller cum network processor which avoids the wired connection between the home router and WPPU, as well as avoiding the cost of an intermediate device between the WPPU and home router.

11.3 THE SCENARIO OF PROPOSED SYSTEM

The proposed simple-link sensor network-based remote monitoring of physiological signals from multiple patients consists of two main sections, the patient's home and the hospital, shown in Fig. 11.3. The home section consists of a WPSU installed at the patient's home which measures the physiological signals and sends these to the remote monitoring station (at the hospital) through a wireless access point unit. The received data of every patient is stored in the database server for review. The precise conditions of every patient are monitored and analyzed by a doctor using a GUI.

11.3.1 Patient Home Module

The patient home module consists of a WPSU and home router. The home router may be the common router used for internet access in the home. So, the main part of the patient home module is WPSU. Since multiple patients are considered, every patient will be made available with WPSU. WPSU consists of an ECG sensor, which can be a ring electrode to sense the electrical activities of the heart, an instrumentation amplifier to amplify the low

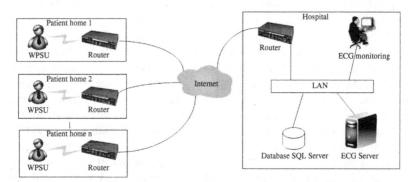

FIGURE 11.3 Scenario of the proposed system.

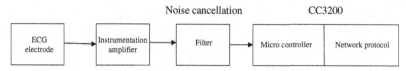

FIGURE 11.4   Block diagram of WPSU.

amplitude ECG signal, a noise cancellation circuit to cancel unwanted signals like muscle, noise, and powerline radiations, and a CC3200 single-chip wireless microcontroller to process the sensed ECG signal and to send the data to the receiver of the access point (home router). From the access point, the data is sent to the hospital via the Internet.

11.3.1.1   Wireless Portable Sensor Unit

The WPSU consists of an instrumentation amplifier, noise cancellation circuit, and a CC3200 single chip microcontroller with IP network module. A block diagram of WPSU is shown in Fig. 11.4.

The patient's body is attached to an ECG sensor to get the ECG signals. The signals from the ECG sensor may be weak, in the range of 0.1−0.3 mV, so that sampling and further processing of the signal are not possible. As a result, an instrumentation amplifier is used to amplify the low amplitude ECG signal to 1 V. The amplified signal is then fed to a noise-cancellation circuit in which all the unwanted signals in the ECG signals are removed. INA2322 IC [24] can be used to perform the operations like ECG signal amplification and noise cancellation. The filtered signal is then given to the microcontroller for further processing and the processed signal reaches the home router through the IP network provided.

CC3200 is a single-chip wireless microcontroller unit which can be used for Internet of Things applications [25]. It is equipped with applications of microcontroller, Wi-Fi, network processor and power management subsystems. It has an arm cortex m4 core microcontroller which works at 80 MHz, a Wi-Fi network processor subsystem with Wi-Fi internet on a chip facility. It is loaded with Wi-Fi and IPs in ROM. Also, the integrated DC−DC converter supports a wide range of supply voltages. A functional block diagram of CC3200 is shown in Fig. 11.5.

11.3.1.2   Wireless Access Point (Home Router)

The home router has the functionalities to provide a wireless access point. The home unit consists of a wireless router to establish a connection with the server in the hospital. In the proposed system, this router will be the common router used in the home to provide an internet facility. Tp-link-WE740N wireless router is used in the proposed system for experimental analysis,

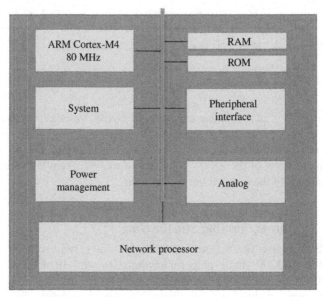

FIGURE 11.5 Functional block diagram of CC3200.

and is compatible with 80211b/g/n devices and has a wireless data rate of 150 mbps.

11.3.2   Hospital Section

The hospital section receives signals continuously from the home by using high-speed internet. So, there is no time lag in sensing, transmission, and processing. The home section consists of a computer server, a windows server is the operating system, and the database is handled by an SQL server. Fig. 11.6 shows the sequence of connection establishment between the WPSU and ECG server. Also, utility software is used to display the real-time ECG signals of various patients on different computers in the hospital for the desired actions. The components of the hospital sections are ECG server, ECG database, and ECG monitoring.

11.3.2.1   Application Software

ECG server application receives the ECG signal from the patient's home via the Internet. It works based on TCP protocols and a real-time client-server, which can be developed using Microsoft development environment. A unique TCP port is assigned to every patient. The primary function of the software is the storage of real-time ECG signals. Using these software and protocols, zero-time lag for the storage and processing of information of multiple patients is possible.

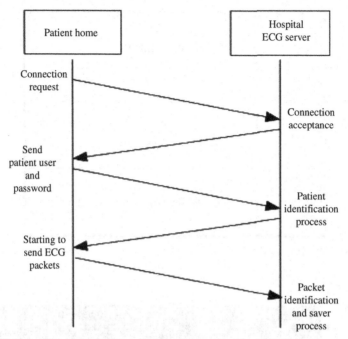

FIGURE 11.6   Communication process between home unit and hospital unit.

11.3.2.2   Electrocardiography Database

The ECG database consists of the patient's table. In the patient's table, personal information, like name, age, date of birth, gender, blood type, address, and disease, are stored. The main structure of the ECG signals' database diagram is shown in Fig. 11.7 By using a "Patient ID," the entire record of a patient can be obtained from the ECG signal table, where 64 samples are taken for every ECG signal recording.

11.3.2.3   Electrocardiography Monitoring

The visualization and analysis of the ECG signals are done by ECG monitoring software. The main functionalities of this software include maintaining the patient database, online plotting, offline plotting, and presentation of the latest recorded data for all active patients. In the patient database, patient information modification and the addition or removal of patients are possible. Online plotting is using to provide a visualized ECG of multiple patients whenever required. Also, all the recent data regarding the active patients (nodes) can be shown in a single window. It has a provision to detect the nodes which are not active due to any kind of problems, such as a communication link disconnect or failure of sensors, etc. Alarms are given to nurses or clinicians to take necessary action to monitor the problem associated with

Patient credentials	
Name	Revathy
Age	23
D.O.B.	22/01/1992
Gender	Female
Blood Type	AB +ve
Address	Block-4/A, Chamar Apartments, Cochin, Kerala
Disease	Arthritis

FIGURE 11.7   ECG signal database.

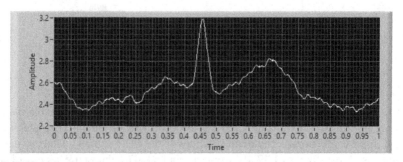

FIGURE 11.8   Extraction of ECG signal.

a particular patient (node). Fig. 11.8 shows the extraction of the ECG signal at the monitoring side.

11.4   RESULTS AND DISCUSSION

11.4.1   Evaluation of Current Consumption of Wireless Portable Sensor Unit

In the proposed system, WPSU mainly has two modes of operation, the active mode and low power mode. Every second, the ECG readings are sent to the remote monitoring system in the hospital section. Each reading consists of 64 samples and are sent to the hospital's server as packets. "567.6×10^{-6}" seconds is required to transmit the data. So, in the remaining time the WPSU will be in a low-power sleep or idle mode. This will increase power efficiency since the device uses batteries. In this section, the average current consumption of WPSU is evaluated.

Sleep mode current contribution of WPSU
$$= \text{(Sensor idle current} + \text{CC3200 idle current)} \times$$
$$\quad \text{(Period of transmission} - \text{Application execution time)}$$
$$= (0.01 \times 10^{-6} \text{ A} + 15.3 \times 10^{-3} \text{ A}) \times (1 \text{ s} - 567.6 \times 10^{-6} \text{ s})$$
$$= 0.01529 \text{ A s}$$

Active mode current contribution of WPSU
$$= \text{(Sensor event current consumed} \times \text{time executed)} +$$
$$\quad \text{(CC3200 event current consumed} \times \text{time executed)}$$
$$= (40 \times 10^{-6} \times 567.6 \times 10^{-6} \text{ A s}) + (278 \times 10^{-6} \times 567.6 \times 10^{-6} \text{ A s})$$
$$= 1.804 \times 10^{-7} \text{ A s}$$

Average current consumption of WPSU
$$= \frac{\begin{array}{c}\text{Sleep current contribution} \\ \text{of WPSU}\end{array} + \begin{array}{c}\text{Active current contribution} \\ \text{of WPSU}\end{array}}{\text{Period of transmission}}$$
$$= \frac{0.01529 \text{ A s} + 1.804 \times 10^{-7} \text{ A s}}{1 \text{ s}}$$
$$= 0.01529 \text{ A}$$

11.5 LIFE EXPECTANCY OF WIRELESS PORTABLE SENSOR UNIT

The battery capacity used in the proposed system is 1 A h. The life expectancy of WPSU can be calculated by considering the theoretical conditions of the battery.

Total days of operation of WPSU
$$= \frac{\text{Current rating of battery}}{\text{Average current consumption}}$$
$$= \frac{1 \text{ A h}}{0.01529 \text{ A}}$$
$$= 65.40 \text{ h}$$
$$\approx 2 \text{ days } 17 \text{ h } 24 \text{ min}$$

11.6 MAXIMUM TRANSMISSION BETWEEN THE WIRELESS PORTABLE SENSOR UNIT AND WIRELESS ACCESS POINT

To calculate the maximum range between the WPSU and the network router, an experiment was conducted as follows, the network router is made

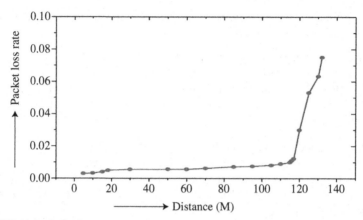

FIGURE 11.9   Packet loss rate versus distance graph.

immovable and WPSU is moved to various distances. To obtain the maximum possible distance between WPSU and the router, 12,500 packets were sent to the server over various distances. Fig. 11.9 shows the packet loss rate versus distance graph. From the graph, it is observed that if the patient is within a range of 119 m, the packet-loss rate is at a minimum, and is around less than 2%.

11.7   CONCLUSION

The remote monitoring of physiological signals has a very good scope in addressing the problems facing patients with chronic diseases. This chapter presents a novel system which uses a wireless sensor network to monitor ECG signals with high reliability and accuracy. Several significant experiments have been performed on this system. Firstly, the life expectancy of WPSU has been estimated from the current consumption of each unit. Secondly, it is observed that the WPSU can maintain the wireless link quality if the distance is less than 119 m line of sight from the wireless access point. Additional mobility can be achieved by increasing the number of access points (home router).

REFERENCES

[1]   N. Fisal, R.A. Rashid, M.A. Sarijari, H.M. Nasir, ECG monitoring system using WSN for home care environment, 2008.

[2]   A.B. Tagad, P.N. Matte, Design and development of wireless remote POC patient monitoring systems, IJEATE 3 (12) (2013) 435–439.

[3]   J. Dong, H.-H. Zhu, Mobile ECG detector through GPRS/Internet, Proceedings 17th IEEE Symposium on Computer-Based Medical Systems, IEEE, Bethesda, MD, USA, 2004.

[4] N.R. Deshmukh, N.S. Nagekar, A.M. Harde, Wireless E.C.G. monitoring on computer using Zigbee technology, Discovery 18 (52) (2014) 58−63.

[5] R. Ramu, S. Kumar, Real time monitoring of ECG using ZigBee technology, IJEAT 3 (2014) 169−172.

[6] Cai Ken, Liang, Development of remote monitoring cardiac patient monitoring system using GPRS, 2010 International Conference on Biomedical Engineering and Computer Science (ICBECS), April 23−25, IEEE, Wuhan, China, 2010.

[7] M. Rajput, S. Pai, U. Mhapankar, Wireless transmission of biomedical parameters using GSM technology, 1, IJESE, 2013, pp. 83−85.

[8] R. Sukanesh, S. Veluchamy, M. Karthikeyan, A portable wireless ecg monitoring system using gsm technique with real time detection of beat abnormalities, IJER 3 (2014) 108−111.

[9] Noureddine, Fethi, Bluetooth portable device for ECG and patient motion monitoring, nature & technology, 2010.

[10] J. Morak, H. Kumpusch, D. Hayn, R. Modre-Osprian, G. Schreier, Design and evaluation of a tele monitoring concept based on NFC-enabled mobile phones and sensor devices, IEEE Trans. Inform. Technol. Biomed. 16 (1) (2010) 17−23.

[11] X.-M. Guo, J. Weng, L.-S. Chen, Z.-H. Yuan, X.-R. Ding, M. Lei, Study on real time monitoring technique for cardiac arrhythmia based on smartphone, J. Med. Biol. Eng. 33 (2013) 394−399.

[12] P. Corbishley, E.R. Villegas, Breathing detection: towards a miniaturized, wearable, battery operated monitoring system, IEEE Trans. Biomed. Eng. 55 (1) (2008) 196−204.

[13] K.A. Ng, P.K. Chan, A CMOS analog front-end IC for portable EEG/ECG monitoring applications, IEEE Trans. Circuit. Syst. 52 (11) (2005) 2335−2347.

[14] Fensli, Gunnarson, Gunderson, A wearable ECG-recording system for continuous arrhythmia monitoring in a wireless tele-home-care situation, 18th IEEE Symposium on Computer-Based Medical Systems (CBMS'05), July 25, IEEE, Dublin, Ireland, 2005.

[15] F. Zhang, Y. Lian, QRS detection based on multiscale mathematical morphology for wearable ECG devices in body area networks, IEEE Trans. Biomed. Circuit. Syst. 3 (4) (2009) 220−228.

[16] S. Saravanan, Remote patient monitoring in tele medicine using computer communication network through Bluetooth, Wi-Fi, Internet Android Mobile, IJARCCE 3 (7) (2014) 7590−7596.

[17] N.G. Prerana, G.S. Nidhi, A.N. Cheeran, Android application for ambulant ECG monitoring, IJARCCE 3 (5) (2014) 6465−6468.

[18] R.S. Deshmukh, Wi-Fi based vital signs monitoring and tracking system for medical parameters, IJETT, vol. 4(5), 2013, pp. 1935−1938.

[19] A.S. Shebi, B.C. Pillai, R. Rajesh, Design and implementation of Wi-Fi based imaging system on concerto platform, Annual International Conference on Emerging Research Areas, 2013.

[20] T. Schrama, Developing a portable wireless physiological monitor using the NI LabVIEW Mobile Modules, National Instruments, and LabVIEW CF-6004, 2011.

[21] F. Touatia, R. Tabish, A.B. Mnaouer, A real time BLE enabled ECG system for remote monitoring, ICBET, 2013.

[22] R.S. Dilmaghani, H. Bobarshad, M. Ghavami, S. Choobkar, C. Wolfe, Wireless sensor networks for monitoring physiological signals of multiple patients", IEEE Transactions on Biomedical Circuits and Systems, vol. 5(4), IEEE, 2011, pp. 347−356.

[23] Takur, Murthy, Wireless Sensor Networks for Monitoring Physiological Signals of Multiple Patients, IJSR, 2013.

[24] Texas Instruments, microPower, Single-Supply, CMOS INSTRUMENTATION AMPLIFIER, SBOS174B datasheet, December 2000 (Revised February 2006).

[25] Texas Instruments, CC3200 SimpleLink™ Wi-Fi® and Internet-of-Things solution, a single-chip wireless MCU, SWAS032F datasheet, July 2013 (Revised February 2015).

Chapter 12

Hybrid Approach for Classification of Electroencephalographic Signals Using Time−Frequency Images With Wavelets and Texture Features

N.J. Sairamya, L. Susmitha, S. Thomas George and M.S.P. Subathra
Karunya Institute of Technology and Sciences, Coimbatore, India

12.1 Introduction	**253**	
12.2 Methodology	**255**	
12.2.1 Dataset	255	
12.2.2 Short-Time Fourier Transform Spectrogram	256	
12.2.3 Discrete Wavelet Transformation	258	
12.2.4 Local Binary Pattern	258	
12.2.5 Local Tetra Pattern	259	
12.2.6 Gray-Level Cooccurrence Matrix	260	
12.2.7 Artificial Neural Network	261	

12.2.8 Support Vector Machine	262
12.2.9 Feature Selection	262
12.2.10 Particle Swarm Optimization	263
12.3 Results and Discussion	**264**
12.3.1 Wavelet Decomposition	265
12.3.2 Feature Extraction	265
12.3.3 Classification Accuracy Results	265
12.4 Conclusion	**271**
References	**272**

12.1 INTRODUCTION

Epilepsy is a calamitous neurological disorder caused by abnormal hyperactivity of neurons in the brain, affecting different parts of the central nervous system [1]. Electroencephalographic (EEG) is the most frequently used method to diagnose epileptic seizure disorders, by measuring the electrical communication between the neurons in the brain and constructing a signal to represent the brain's electrical activity [2]. The conventional diagnostic approach depends on a skilled neurophysiologist to visually examine the long (protracted) EEG signal, which is tedious, time consuming, and unreliable. Hence, an automated diagnostic tool to assist neurophysiologists in diagnosing epileptic seizure disorders is in high demand.

Feature extraction and classification are the two main parts that are associated with the automated diagnosis of epileptic seizure disorders. During feature extraction, the EEG signals are generally characterized in frequency [3], time [4] or t−f domain [5−7], and various features are extracted from it for detecting epileptic seizure activities. Recently, image processing methods like the local binary pattern (LBP) [1], local neighbor descriptive pattern (LNDP), and local gradient pattern technique (LGP) [8] are applied to one-dimensional EEG signals for better classification of epileptic and nonepileptic signals. Various methods based on t−f representation of EEG signals are also proposed ([9−11,31]).

Haralick features were extracted from the t−f image to classify seizure signals using a support vector machine and has obtained an accuracy rate of 99.125% [31]. In Ref. [11], the researchers combined the t−f image features with the features obtained from the t−f signal to enhance the performance of automatic epileptic detection. Texture pattern techniques, such as the gray-level cooccurrence matrix (GLCM), LBP, and texture feature coding method (TFCM), were proposed to analyze the t−f images obtained using spectrogram of short-time Fourier transform (STFT) [12]. Classification of epileptic signals were carried out by using the support vector machine (SVM) technique and obtained a classification accuracy of 100%. The t−f image obtained using quadratic time−frequency distribution was analyzed using the LBP technique and obtained an accuracy of 99.3% in classifying epileptic and nonepileptic signals using SVM classifier [13].

The t−f image represents the energy distribution of nonstationary EEG signals in a two-dimensional space. The energy distribution of epileptic and nonepileptic EEG signals in different frequency bands consists of inequitable texture patterns. Hence, in Ref. [12], the researchers proposed LBP texture features with a higher dimension, which is suitable for characterizing the t−f image of epileptic and nonepileptic signals. Recently a novel feature extraction method to achieve a high-resolution spectral estimation was proposed using multiscale radial basis functions (MRBF) and modified particle swarm optimization (MPSO) [14]. To improvise the radial bias function in real-time applications Fathi and Montazer [15] proposed particle swarm optimization (PSO) to optimize the Optimum Steepest Decent algorithm. A modified PSO algorithm was suggested by Ge et al. [16] to train an artificial neural network (ANN).

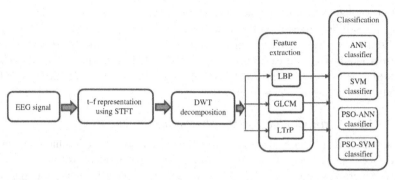

FIGURE 12.1   Block diagram of the proposed system.

In the proposed study, the t−f image representation of each EEG signal is acquired using the STFT. The wavelet transform (WT) is applied to each t−f image which results in horizontal (H), diagonal (D), and vertical (V) components of the t−f image (Fig. 12.1). Effective features with high discriminative power are extracted from each component using texture pattern techniques, such as LBP, local tetra pattern (LTrP), and GLCM. ANN and SVM classifiers are used to categorize epileptic and nonepileptic signals using the extracted features. Further, for feature selection, PSO is proposed to improvise the classification performance. The classification performance is evaluated based on the statistical measures.

The rest of this chapter is organized as follows: Methodology and the related theories are explained in Section 12.2. Experimental results and a discussion are provided in Section 12.3. In Section 12.4 the conclusions are drawn.

12.2   METHODOLOGY

A brief discussion on the EEG dataset, t−f representation of EEG signal, wavelet decomposition of the images, LBP, LTrP, GLCM, SVM, ANN, PSO, and feature selection based on PSO, are provided in this section.

12.2.1   Dataset

12.2.1.1   University of Bonn Dataset
In this chapter, a standard and a widely used EEG dataset contributed by the University of Bonn (UB) is used for evaluating the proposed system. The dataset consists of 500 single-channel EEG signals and each signal belongs to one of the following groups: A (healthy-eyes open), B (healthy-eyes closed), C (inter ictal-hippocampal formation of opposite hemisphere), D (inter ictal-epileptogenic zone), and E (seizure). The EEG signals in A and B

are regarded as healthy signals. The EEG signals in C and D are seizure-free signals, while E represents the signals with seizure epochs. All the EEG signals last for 23.6 seconds with a sampling rate of 173.61 Hz. Hence, each signal length is given as $23.6 \times 173.6 = 4097$, with a bandwidth of 86.8 Hz. Further details on the EEG dataset can be found in Ref. [17]. EEG signals from each group is shown in Fig. 12.2.

12.2.1.2 Karunya Electroencephalographic Database

The EEG dataset contributed by the Karunya Institute of Technology and Sciences is used for evaluating the proposed system in classifying normal and generalized epilepsy. The 16-channel EEG signals from the patients is measured based on 10−20 electrode placements (standard international system) from neonate, infant, and adults using a bipolar setting. For EEG recording, the patient's scalp is carefully prepared to allow a contact impedance of less than 5 kΩ and the signals are acquired with a sampling rate of 256 Hz using analog pass band ranges from 0.01 Hz to 100 Hz. Each EEG segment consists of 2560 sampling points; hence in the order of milliseconds, epileptic seizure activities are captured with high-temporal resolution. The EEG dataset of normal and generalized epileptic patients accumulated from a diagnostic center in Coimbatore, Southern India, is used in this research [18,19].

12.2.2   Short-Time Fourier Transform Spectrogram

A spectrogram of STFT represents the normalized, squared magnitude of STFT coefficients [12]. Hence, the energy in the t−f signal is equal to the

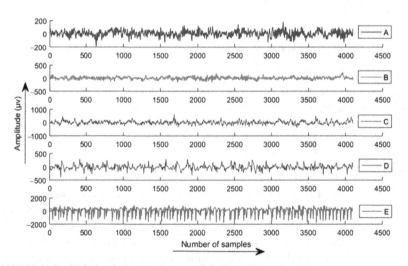

FIGURE 12.2   EEG signals from each group of the Bonn University database.

energy in the spectrogram of STFT. In STFT, the time domain signals are divided into smaller parts (window) and Fourier transform is computed for each windowed section to obtain the frequencies. The window location is slid through the entire data to obtain STFT coefficients. Based on [12], STFT is mathematically presented as:

$$X(n, \omega) = \sum_{m=-\infty}^{\infty} x(m)w(n-m)e^{-j\omega n} \qquad (12.1)$$

where $X(n)w(n-m)$ is a short-time part of the input signal $x(n)$ at time n. In addition, a discrete STFT is defined as:

$$X(n, k) = X(n, \omega)|_{\omega=(2\pi k/N)} \qquad (12.2)$$

where N shows the number of discrete frequencies. Thus the spectrogram in logarithmic scale is defined as:

$$S(n, k) = \log|X(n, k)|^2 \qquad (12.3)$$

The spectrogram of the EEG signals from group A, D, and E of the UB dataset are shown in Fig. 12.3.

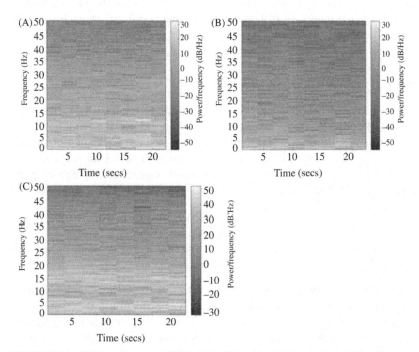

FIGURE 12.3   STFT spectrogram of EEG signal (A) healthy, (B) interictal, and (C) epileptic.

12.2.3 Discrete Wavelet Transformation

In recent years, WT is developing into an essential tool in the analysis of nonstationary signals and feature extraction. Mathematically, a wavelet resembles a wave-like oscillation which is time-bound and assists in capturing the frequency and temporal information of a signal. A smaller number of scales with various amounts of translations at each scale is used to analyze a signal in discrete wavelet transform (DWT). The critical sampling of continuous WT with a scaling parameter of $a = 2^{-j}$ produces the DWT [20] and is given as:

$$W(j \cdot k) = \int_t f(t) 2^{j/2} \psi(2^j t - k) dt \qquad (12.4)$$

where j indicates the set of discrete translations and k represents the dilations [21]. The DWT on an image consists of one-dimensional, row-column filtering of the original image which results in four image components, namely the low-low (LL), low-high (LH), high-low (HL), and high-high (HH) detail images, which are a decimated version of the original image by two for every level of decomposition. The LH and HL images capture the horizontal and vertical information, respectively, whereas the LL image is a low-resolution decimated version of the original and HH contains the diagonal information [22]. This work analyzes the suitability of the coiflet wavelet family, among others, in EEG spectrogram image processing for better classification of epileptic disorders.

12.2.4 Local Binary Pattern

LBP is an effective texture pattern descriptor introduced by Ojala et al. [23] to describe the local texture patterns of an image. It is widely used in the applications based on image processing. The LBP works in a block size of 3×3, in which the center pixel is used as a threshold for the neighboring pixel, and the LBP code of a center pixel is generated by encoding the computed threshold value into a decimal value. Mathematical expression of LBP is given as:

$$\begin{aligned} \text{LBP} &= \sum_{i=0}^{P-1} s(n_i - G_c) 2^i \\ s(x) &= \begin{cases} 1, & if \ x > 0 \\ 0, otherwise \end{cases} \end{aligned} \qquad (12.5)$$

where P is the number of neighborhood pixels, n_i represents the ith neighboring pixel, and c represents the center pixel. The histogram features of size 2^P is extracted from the obtained LBP code. Hence, for eight neighboring pixels the histogram feature vector length of 256 is obtained. The LBP operation is shown in Fig. 12.4 with a G_c value of 10 and eight neighboring pixels.

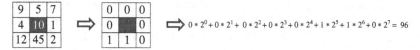

FIGURE 12.4   LBP operation.

12.2.5 Local Tetra Pattern

The pattern techniques, like LBP, obtain the texture pattern features of the query image based on the distribution of edges, which are coded only in two directions (positive or negative). The performance of the LBP method can be improved by differentiating the edges in more than two directions. Therefore LTrP was introduced by Murala et al. [24] to extract the texture pattern features more effectively based on four directions. LTrP describes the spatial structure of the local texture using the direction of the center pixel G_c.

Consider an image as I, the first-order derivative along the direction 0 and 90 degrees are denoted as $I_\theta^1(G_p)|_{\theta=0\,\mathrm{deg},90\,\mathrm{deg}}$. Let G_c be the center pixel of query image I, and G_h and G_v the horizontal and vertical neighborhood pixels of the center pixel G_c, respectively. The first-order derivatives at the center pixel are given as:

$$I_{0\,\mathrm{deg}}^1(G_c) = I(G_h) - I(G_c) \tag{12.6}$$

$$I_{90\,\mathrm{deg}}^1(G_c) = I(G_h) - I(G_c) \tag{12.7}$$

and the respective directions of the center is computed as

$$I_{\mathrm{Dir.}}^1(G_c) = \begin{cases} 1, & I_{0\,\mathrm{deg}}^1(G_c) \geq 0 \text{ and } I_{90\,\mathrm{deg}}^1 \geq 0 \\ 2, & I_{0\,\mathrm{deg}}^1(G_c) < 0 \text{ and } I_{90\,\mathrm{deg}}^1 \geq 0 \\ 3, & I_{0\,\mathrm{deg}}^1(G_c) < 0 \text{ and } I_{90\,\mathrm{deg}}^1 < 0 \\ 4, & I_{0\,\mathrm{deg}}^1(G_c) \geq 0 \text{ and } I_{90\,\mathrm{deg}}^1 < 0 \end{cases} \tag{12.8}$$

From the above equation, the four possible directions of each pixel is computed which can be either 1, 2, 3, or 4 and then the given image is converted into either of the four values, that is, directions. Finally, the second-order derivative is defined as:

$$\mathrm{LTrP}^2(G_c) = \{f_1(I_{\mathrm{Dir.}}^1(G_c), I_{\mathrm{Dir.}}^1(G_1)), f_1(I_{\mathrm{Dir.}}^1(G_c), I_{\mathrm{Dir.}}^1(G_2)), \ldots, f_1(I_{\mathrm{Dir.}}^1(G_c), I_{\mathrm{Dir.}}^1(G_P))\}|_{P=8} \tag{12.9}$$

$$f_1(I_{\mathrm{Dir.}}^1(G_c), I_{\mathrm{Dir.}}^1(G_P)) = \begin{cases} 0, & I_{\mathrm{Dir.}}^1(G_c) = I_{\mathrm{Dir.}}^1(G_p) \\ I_{\mathrm{Dir.}}^1(G_p), & otherwise \end{cases} \tag{12.10}$$

From Eqs. (12.9) and (12.10), an 8-bit tetra pattern for each pixel is obtained. Then the obtained patterns are separated into four parts based on the direction of the center pixel. Finally, the tetra patterns for each part (direction) are transformed into three binary patterns. Let the direction of the center pixel $I^1_{\text{Dir.}}(G_c)$ obtained using Eq. (12.10) be "1"; then, LTrP^2 is defined by separating it into three patterns which is given as follows:

$$\text{LTrP}^2\big|_{\text{Direction}=2,3,4} = \sum_{p=1}^{P} 2^{(p-1)} \times f_2(\text{LTrP}^2(G_c)\big|_{\text{Direction}=\varnothing} \qquad (12.11)$$

$$f_2\left(\text{LTrP}^2(G_c)\right)\big|_{\text{Direction}=\varnothing} = \begin{cases} 1, & \text{if } \text{LTrP}^2(G_c) = \varnothing \\ 0, & \text{else} \end{cases} \text{ where } \varnothing = 2,3,4$$
$$(12.12)$$

Similarly, the other three patterns for the remaining three directions of the center pixels are converted into corresponding binary patterns. Hence 12 binary patterns (4 direction $\times$ 3 patterns for each direction) are obtained.

Guo et al. [25] proposed the magnitude LBP, using the magnitude component of the local difference operator, along with sign LBP for texture pattern classification. Hence, this concept is used to propose the thirteenth binary pattern by using the magnitude of horizontal and vertical first-order derivatives using:

$$M_{I^1(G_p)} = \sqrt{\left(I^1_{0\,\text{deg}}(G_p)\right)^2 + \left(I^1_{90\,\text{deg}}(G_p)\right)^2} \qquad (12.13)$$

$$\text{LP} = \sum_{p=1}^{P} 2^{(p-1)} \times f_1\left(M_{I^1(G_p)} - M_{I^1(G_c)}\right)\big|_{P=8} \qquad (12.14)$$

Finally, a total of 13 binary patterns (12 directional binary patterns + 1 magnitude bit) is achieved. Therefore it is observed that that the second-order LTrP operator is able to obtain more detailed directional information compared to LBP operations. Each binary pattern is converted into a LTrP code and the corresponding histogram features are evaluated. All 13 binary patterns are concatenated into a single feature vector to form the LTrP features. The LTrP operation is shown in Fig. 12.5.

12.2.6 Gray-Level Cooccurrence Matrix

According to Haralick et al. [26], in GLCM the relative frequency $p(i,j)$ is computed by considering the distribution of two-pixel intensities, i and j, in a given direction, θ, at a distance, d. The number of gray levels (G) considered in an image is always equal to the number of rows and columns in GLCM, which is sensitive to the number of gray levels chosen. Hence, to have a minimum computational level, the G value should be small. The GLCM for four different orientations (0, 45, 90, and 135 degrees) at a

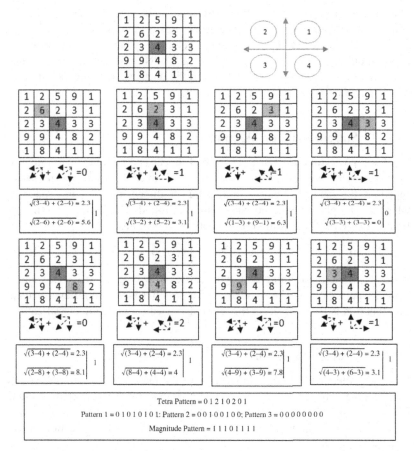

FIGURE 12.5 LTrP operation for direction 1 with the center pixel 4 and eight neighborhood pixels.

distance of 1, with $G = 8$ is computed and the 22 different features provided by Abdel-Nasser et al. [27] are extracted from each GLCM. The 22 features from each GLCM is concatenated to form a single feature vector with 88 dimensions to classify epileptic brain maps from artifact brain maps.

12.2.7 Artificial Neural Network

ANN architecture is based on the structure and function of the biological neural network. Similar to neurons in the brain, ANN also consists of neurons which are arranged in various layers. Feed forward neural network is a popular neural network which consists of an input layer to receive the external data to perform pattern recognition, an output layer which gives the problem solution, and a hidden layer is an intermediate layer which separates the other layers. The adjacent neurons from the input layer to output layer are connected through acyclic arcs. The ANN uses a training algorithm to learn the datasets which modifies the neuron weights depending on the error rate between target and actual output. In general, ANN uses the back propagation algorithm as a training algorithm to learn the datasets. The general structure of ANN is shown in Fig. 12.6.

12.2.8   Support Vector Machine

SVM is a powerful classifier used for both linear and nonlinear classification by changing the kernel functions utilized [28]. In SVM, by using the kernel functions the data are mapped into a higher dimensional feature space, in which a hyperplane separating the classes are found. The general solution for finding the hyperplane with kernel functions is given as:

$$f(x) = \sum_{i} a_i y_i K(x_i, x) \tag{12.15}$$

where $\{x_i, y_i\}$ is the training data with classes y_i belonging to $\{-1, 1\}$ and K is the kernel function. The standard SVM classifier with a multilayer perceptron function is first trained in this work and used for epileptic classification.

12.2.9   Feature Selection

In feature selection, the high dimensions of the feature vector are reduced by selecting the optimum subsets of the entire feature set. The selected optimum features consist of the essential information needed for the classification task with a reduced computational burden. In this study, feature selection is carried out using PSO.

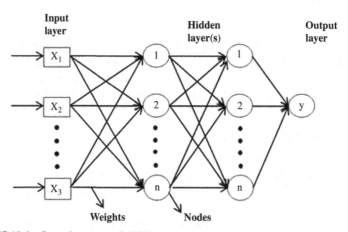

FIGURE 12.6   General structure of ANN.

12.2.10   Particle Swarm Optimization

A PSO algorithm is used for locating optimal values of continuous nonlinear functions, and was introduced by Kennedy and Heberhart [29]. PSO is an evolutionary computation technique inspired by social behaviors, such as swarming theory, bird flocking, and fish schooling [30]. It is based on the principle that each solution in this algorithm is represented as a vector, called a particle. Each particle is assigned a position in the search space to examine optimum solutions. Hence, each particle has a velocity. During each movement, the velocity and position of each particle are updated based on its self-experience and its neighbors. The best previous position of the particle is recorded as the personal best (*pbest*), and the best position obtained by the population thus far is called (*gbest*). Based on *pbest* and *gbest*, PSO searches for optimal solutions by updating the velocity and the position of each particle. The detailed algorithm is:

Step 1: The constant values of cognitive and social acceleration factors (c_1,c_2), inertia weight factor (w), maximum iteration (k_{max}), and random numbers (r_1,r_2) are initialized.
   Randomly initialize particle positions $x_0(i)$ for $i = 1, 2,\ldots\ldots\ldots, p$.
   Randomly initialize particle velocities $v_0(i)$ for $i = 1, 2\ldots\ldots\ldots, p$.
Step 2: Assign $m = 1$.
Step 3: Function value f_m is evaluated using the design space coordinates $x_m(i)$

$$pbest(i) = x_m(i), \text{ iff}_m \geq f_{pbest}$$

$$gbest = x_m(i), \text{ iff}_m \geq f_{gbest}$$

Step 4: The particle velocity is updated using:

$$V_{m+1(i)} = w \times (v_{m(i)}) + c_1 r_1 \times (pbest_{m(i)} - x_{m(i)}) + c_2 r_2 (gbest_m - x_{m(i)})$$

$$(12.16)$$

The particle position vector is updated using:

$$X_{m+1(i)} = x_{m(i)} + v_{m+1(i)} \qquad (12.17)$$

Step 5: The variable i is incremented. If $i > p$, k is incremented and i is set as 1.
Step 6: The steps 3−5 are repeated until k_{max} is reached.

The algorithm stops when a predefined criterion (good fitness value or k_{max}) is met. The genetic optimization is initiated by a population of 20 chromosomes. The initial coding for each particle is randomly generated. For each individual, a suitable fitness function is estimated. The fittest individuals are selected and the new population is generated using the crossover and mutation operations. This process is continued for a particular number of generations and the fitness function (18) is used to calculate the fittest chromosome. The fitness function (18), minimizes the classification error rate attained by the selected features during the evolutionary training process

$$F1 = \text{Error rate} = \frac{FP + FN}{TP + TN + FP + FN} \tag{12.18}$$

where true negative (TN) represents the number of seizure signals detected as normal signals, true positive (TP) represents the number of seizure signals identified correctly, false positive (FP) indicates the number of nonseizure signals detected as seizure signals, and false negative represents the number of seizure signals detected as nonseizure signals. A particle in PSO represents a vector of n real numbers, where n is the total number of data in the feature set and also represents the dimensionality of the search space. If position vector $(x_{id}) > \theta$, the feature d is selected or else, when not selected, the θ is the threshold value.

12.3 RESULTS AND DISCUSSION

The experiments are run on Intel core i3-7100U (2.40 GHz) processor with 4 GB RAM using MATLAB (Math Work R2016b) software. The experiments are conducted on two different datasets, namely the UB database and Karunya EEG database. In the UB database, 500 single-channel EEG signals from all five groups are considered for evaluating the proposed system with the following classification problems: AB-E (Normal and epileptic), CD-E (interictal and epileptic), and ABCD-E (nonseizure and seizure). From the Karunya EEG database, five normal and generalized EEG datasets are considered to analyze the performance of the proposed algorithm in classifying the normal and generalized EEG signals. The performance of the proposed system is tested using all the four classifiers, and each classification problem is evaluated using the confusion matrix with the quantity of testing data incorrectly classified and correctly classified. From each database, 60% of the dataset is used as the training set and 40% of the dataset is used as the testing set. The following statistical measures are used to evaluate the performance of the proposed system,

1. $\quad \text{Accuracy} = \dfrac{TP + TN}{TP + FP + TN + FN} \qquad (12.19)$

2. $\quad \text{Sensitivity} = \dfrac{TP}{TP + FN} \qquad (12.20)$

3. $\quad \text{Specificity} = \dfrac{TN}{TN + FP} \qquad (12.21)$

4. $\quad \text{Positive predictive value} = \dfrac{TP}{TP + FP} \qquad (12.22)$

5. $\quad \text{Negative predictive value} = \dfrac{TN}{TN + FN} \qquad (12.23)$

12.3.1   Wavelet Decomposition

Wavelet decomposition is applied to each t−f image representation of the EEG signals resulting in diagonal (D), vertical (V), and the horizontal (H) components which are stored as images and are employed for feature extraction. Seven different wavelets are used in this work to perform the wavelet analysis on each image, namely Daubchis 4, discrete Meyer, Symlet 6, Coiflet 3, Haar, Biorthogonal 5.5, and reverse biorthogonal 6.8. Based on Ref. [20], the decomposition level for each wavelet family is set to 1. Compared to other wavelet families the coiflet wavelets clearly show the patterns of the t−f image in H, V, and D images rather than the other wavelets, respectively.

12.3.2   Feature Extraction

The LBP, GLCM, and LTrP features are extracted from each wavelet's filtered images with different feature vector sets. For LBP a feature vector of 768 (256 features × 3 components), for GLCM 264 feature vector (22 features × 4 orientations × 3 components), and for LTrP 9984 features (3328 features × 3 components) are computed from the wavelet-filtered images. These features are fed into classifiers to analyze the performance of the proposed system.

12.3.3   Classification Accuracy Results

All the four classifiers are tested in this work, in which two of the classifiers use the complete feature set and two of the classifiers with PSO use the optimal feature set. The classification of three different classification problems using the UB dataset with four different classifiers is listed in Table 12.1. A comparison of each feature extraction technique with different classifiers is depicted in Fig. 12.7.

From Fig. 12.7, it is clear that in all three feature extraction techniques, PSO-ANN attained the highest classification accuracy with the optimal feature set. Even with LBP features, ANN attained a maximum accuracy in all cases where the amount of the feature vector required was 768. By using PSO-based ANN a maximum accuracy of 100% is achieved with an optimal feature set of size 60. In GLCM and LTrP feature sets, a significant improvement in the classification accuracy is noted by using PSO-based ANN, as the insignificant features are eliminated.

A comparison of classification accuracy using PSO-ANN with GLCM, LBP, and LTrP features are illustrated in Fig. 12.8. Compared to GLCM and LTrP, the LBP feature attained the highest classification accuracy. Comparison of classification accuracy with different wavelets using LBP features are given in Fig. 12.9.

The performance of the proposed system in classifying N-G experimental cases using the Karunya EEG database is given in Table 12.2 and it is

TABLE 12.1 Classification Accuracy of Three Different Classification Problems in the UB Dataset Using Four Classifiers With GLCM, LBP, and LTrP Feature Sets

S. No	Case	Feature Extraction Technique	SVM (%) (Feature Size)	ANN (%) (Feature Size)	PSO-SVM (%) (Feature Size)	PSO-ANN (%) (Feature Size)
1	AB-E	GLCM	53.33 (264)	92 (264)	63.33 (60)	93.66 (60)
		LBP	97.5 (768)	100 (768)	100 (60)	100 (60)
		LTrP	62.5 (9984)	86 (9984)	71.66 (60)	92.33 (60)
2	CD-E	GLCM	67.5 (264)	90.33 (264)	72.5 (60)	93.66 (60)
		LBP	100 (768)	100 (768)	100 (60)	100 (60)
		LTrP	47.5 (9984)	85(9984)	76.66 (60)	92 (60)
3	ABCD-E	GLCM	57.5 (264)	91.6 (264)	62.5 (60)	94.6 (60)
		LBP	91.5 (768)	100 (768)	99 (60)	100 (60)
		LTrP	42.5 (9984)	91.6 (9984)	68 (60)	93.8 (60)

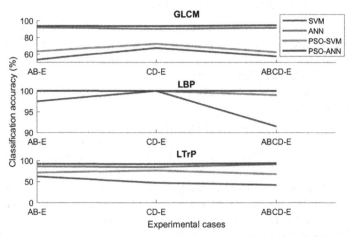

FIGURE 12.7 Comparison of classification accuracy of four different classifiers for three cases using the UB dataset with the feature extraction techniques GLCM, LBP, and LTrP.

depicted in Fig. 12.10. Compared to all the classifiers, PSO-ANN attained the highest classification accuracy with the LTrP features. By using PSO-ANN, a significant improvement in the classification accuracy is noticed as the insignificant features are eliminated.

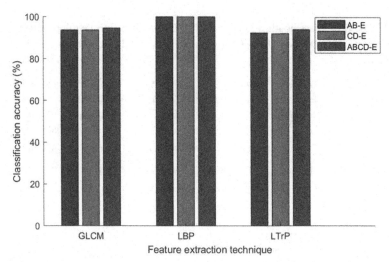

FIGURE 12.8 Comparison of classification accuracy using the PSO-ANN classifier for three cases using the UB dataset with feature extraction techniques GLCM, LBP, and LTrP.

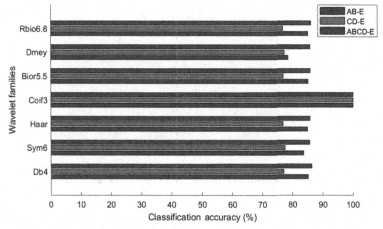

FIGURE 12.9 Comparison of classification accuracy of the PSO-ANN classifier for three cases using the UB dataset with the feature extraction technique LBP for different wavelet families.

Compared to the SVM, ANN, and PSO-SVM classification technique, PSO-ANN achieved maximum classification accuracy in all the experimental cases for the UB and Karunya datasets (Figs. 12.6 and 12.9) with a feature vector length of 60. PSO-ANN classification performance using LBP and LTrP feature extraction techniques for all the experimental cases is listed in Table 12.3. In both databases, the classification accuracy attained by employing the GLCM feature set is less compared to other feature extraction

TABLE 12.2 Classification Accuracy of N-G Classification Problem in the Karunya EEG Dataset Using Four Classifiers With GLCM, LBP, and LTrP Feature Sets

S. No	Case	Feature Extraction Technique	SVM (%) (Feature Size)	ANN (%) (Feature Size)	PSO-SVM (%) (Feature Size)	PSO-ANN (%) (Feature Size)
1	N_G	GLCM	48.43 (264)	86.87 (264)	67.18 (60)	86.87 (60)
		LBP	51.56 (768)	84.37 (768)	70.31 (60)	90.62 (60)
		LTrP	100.00 (9984)	99.37 (9984)	100.00 (60)	100.00 (60)

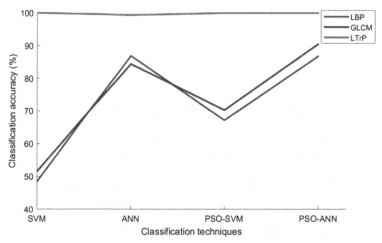

FIGURE 12.10 Comparison of classification accuracy obtained using four different classifiers in classifying normal and generalized epilepsy by using GLCM, LBP, and LTrP features.

techniques. In terms of the UB dataset, the LBP achieved a maximum classification accuracy of 100%, but in the Karunya dataset an accuracy of 51.56% is achieved for identifying generalized epilepsy. LTrP attained a classification accuracy of 92.33% in the UB dataset for classifying the normal and focal (AB-E) EEG dataset, whereas a maximum accuracy of 100% is achieved in the Karunya dataset for classifying the normal and generalized epileptic signals. Obtained results indicate that the LBP feature extraction technique is effective in identifying focal epilepsy, which affects a particular region of the brain, whereas detection of generalized epilepsy affecting all parts of the brain is not effective with the LBP features. Hence, the improved version of LBP, called the LTrP method, is utilized in this work for the

TABLE 12.3 PSO-ANN Classification Performance for All the Classification Problems in the UB and Karunya Databases Using the LBP and LTrP Feature Extraction Method

S. No	Case	Feature Extraction Technique	Accuracy (%)	Sensitivity (%)	Specificity (%)	PPV (%)	NPV (%)
1	AB-E		100	100	100	100	100
2	CD-E	LBP	100	100	100	100	100
3	ABCD-E		100	100	100	100	100
4	N-G		90.62	88.23	93.33	93.75	87.5
5	AB-E		92.33	94.02	88.88	94.5	88
6	CD-E	LTrP	92	91.9	92.22	96.5	83
7	ABCD-E		93.8	96.24	84.15	96	85
8	N-G		100	100	100	100	100

detection of generalized epilepsy with features extracted from four directions of a center pixel.

A fair comparison of the proposed techniques with other methods proposed in the literature are carried out with following conditions: (1) the existing methods are evaluated with the same dataset; (2) signals are transformed into a t–f image; and (3) all the cases mentioned in the literature are evaluated in this work (Table 12.4). From Table 12.4 it is clear that the proposed

TABLE 12.4 Comparison of the Proposed Method With Existing Techniques for the UB Dataset Using t–f Images of the EEG Signal

Author	Method			Accuracy (%)
	t–f Image	Feature	Classification	
			CASE A-E	
[14]	MRBF-MPSO	MRBF-MPSO-PCA	SVM	100
[13]	QTFD	LBP	SVM	99.33
[12]	STFT	LBP		100
		GLCM	SVM	92.5-100
		TFCM		87
[10]	HHT	Histogram features	SVM	99.125
[31]	QTFD	GLCM	SVM	99.125
[9]	QTFD	t–f signal and t–f image features are concatenated	SVM	95.33
Proposed method	STFT	DWT + GLCM	PSO-ANN	93.66
		DWT + LBP		100
		DWT + LTrP		92.33
			CASE CD-E	
[14]	MRBF-MPSO	MRBF-MPSO-PCA	SVM	98.73
Proposed method	STFT	DWT + GLCM	PSO-ANN	93.66
		DWT + LBP		100
		DWT + LTrP		92

technique outperforms the existing technique with the highest classification accuracy of 100% in both the cases. The histogram features obtained from the grayscale image attained an accuracy of 99.125% and 95.33% ([9,10]), which is comparatively less than the proposed method. It is evident that the maximum accuracy of 100% for classifying epileptic and nonepileptic signals is achieved by using the LBP and GLCM method [12]. But the proposed method outruns the existing method in terms of a lesser feature set with high classification accuracy.

Experimental results presented in this section indicate that the proposed method can be employed for effective EEG-based computer aided diagnoses of epilepsy. Compared to GLCM and LTrP methods, the LBP method attained the highest classification accuracy in classifying the normal and focal epileptic signals using the UB dataset. The LTrP method attained the maximum classification accuracy in classifying the generalized and normal EEG signals using the Karunya EEG dataset. Further features extracted from wavelet segmented images improvise the classification accuracy, as the minute details from the t−f images are also computed. Compared to ANN, SVM, and PSO-SVM, the PSO-ANN classifier attained maximum classification accuracy with the reduced feature vector set. The EEG dataset size evaluated in this work is limited; hence, in future, the performance of the proposed system will be evaluated with a larger dataset. The computational complexity of the proposed work is slightly higher than the existing approach, as the t−f images are decomposed into three levels using WT.

12.4 CONCLUSION

A method for automatic diagnosis of epilepsy using STFT for the transformation of EEG signals into t−f images and feature extractions based on texture pattern techniques from the decomposed images are proposed in this research work. The t−f image of the EEG signal is obtained using STFT and the attained t−f image is decomposed into horizontal, diagonal, and vertical components using coiflet WT. From the three components, the most discriminate features are extracted from the wavelet decomposed t−f image using GLCM, LTrP, and LBP methods with a feature vector length of 264, 9984, and 768. Moreover, to improvise the accuracy by reducing the feature set, a PSO-based feature selection method is employed in this work using a SVM and ANN classifier. The extracted features are fed into SVM, ANN, PSO-SVM, and PSO-ANN classifiers for detecting epileptic and nonepileptic signals. The performance of the proposed method is evaluated on two datasets for the following classification problems: normal-seizure (AB-E), interictal-seizure (CD-E), nonseizure−seizure (ABCD-E), and normal-generalized (N-G). The performance of the proposed method has been compared with the existing methods for classification of three classification problems in terms of classification accuracy on the UB dataset. An accuracy rate of 100% is achieved for the cases of AB-E, CD-E, and ABCD-E using LBP-based texture features with a PSO-ANN classifier. For classification of N-G EEG signals, the LTrP feature extraction method achieved the highest classification accuracy. In addition, the proposed approach outperforms the recently proposed methodologies for epileptic detection using t−f representation of EEG signals. In the future, the proposed methodology for EEG data with more channels will be implemented and the reliability of the proposed method will be evaluated.

REFERENCES

[1] Y. Kaya, M. Uyar, R. Tekin, S. Yıldırım, 1D-local binary pattern based feature extraction for classification of epileptic EEG signals, Appl. Math. Comput. 243 (2014) 209−219.

[2] S. Noachtar, J. Rémi, The role of EEG in epilepsy: a critical review, Epilepsy Behav. 15 (1) (2009) 22−33.

[3] V. Srinivasan, C. Eswaran, A.N. Sriraam, Artificial neural network based epileptic detection using time-domain and frequency-domain features, J. Med. Syst. 29 (6) (2005) 647−660.

[4] V. Joshi, R.B. Pachori, A. Vijesh, Classification of ictal and seizure-free EEG signals using fractional linear prediction, Biomed. Signal Process. Control 9 (2014) 1−5.

[5] R. Upadhyay, P. Padhy, P. Kankar, A comparative study of feature ranking techniques for epileptic seizure detection using wavelet transform, Comput. Electr. Eng. 53 (2016) 163−176.

[6] F. Riaz, A. Hassan, S. Rehman, I.K. Niazi, K. Dremstrup, EMD-based temporal and spectral features for the classification of EEG signals using supervised learning, IEEE Trans. Neural Syst. Rehabil. Eng. 24 (1) (2016) 28−35.

[7] M. Peker, B. Sen, D. Delen, A novel method for automated diagnosis of epilepsy using complex-valued classifiers, IEEE J. Biomed. Health Inform. 20 (1) (2016) 108−118.

[8] A.K. Jaiswal, H. Banka, Local pattern transformation based feature extraction techniques for classification of epileptic EEG signals, Biomed. Signal Process. Control 34 (2017) 81−92.

[9] B. Boashash, L. Boubchir, G. Azemi, A methodology for time-frequency image processing applied to the classification of non-stationary multichannel signals using instantaneous frequency descriptors with application to newborn EEG signals, EURASIP J. Adv. Signal Process. 2012 (1) (2012) 117.

[10] K. Fu, J. Qu, Y. Chai, Y. Dong, Classification of seizure based on the time-frequency image of EEG signals using HHT and SVM, Biomed. Signal Process. Control 13 (2014) 15−22.

[11] V. Bajaj, R.B. Pachori, Automatic classification of sleep stages based on the time-frequency image of EEG signals, Comput. Methods Progr. Biomed. 112 (3) (2013) 320−328.

[12] A. Şengür, Y. Guo, Y. Akbulut, Time−frequency texture descriptors of EEG signals for efficient detection of epileptic seizure, Brain Inform. 3 (2) (2016) 101−108.

[13] L. Boubchir, S. Al-Maadeed, A. Bouridane, A.A. Chérif, Classification of EEG signals for detection of epileptic seizure activities based on LBP descriptor of time-frequency images, 2015 IEEE International Conference on Image Processing (ICIP), IEEE, Quebec City, QC, 2015, pp. 3758−3762.

[14] Y. Li, X. Wang, L. Luo, K. Li, X. Yang, Q. Guo, Epileptic seizure classification of EEGs using time-frequency analysis based multiscale radial basis functions, IEEE J. Biomed. Health Inf. 22 (2) (2017). p. 11.

[15] V. Fathi, G.A. Montazer, An improvement in RBF learning algorithm based on PSO for real time applications, Neurocomputing 111 (2013) 169–176.

[16] H. Ge, F. Qian, Y. Liang, W. Du, L. Wang, Identification and control of nonlinear systems by a dissimilation particle swarm optimization-based Elman neural network, Nonlinear Anal. Real World Appl. 9 (4) (2008) 1345–1360.

[17] R.G. Andrzejak, K. Lehnertz, C. Rieke, F. Mormann, P. David, C.E. Elger, Indications of nonlinear deterministic and finite dimensional structures in time series of brain electrical activity: dependence on recording region and brain state, Phys. Rev. E 64 (2001) 061907.

[18] S.E. Selvan, S.T. George, R. Balakrishnan, Range-based ICA using a nonsmooth quasi-newton optimizer for electroencephalographic source localization in focal epilepsy, Neural Comput. 27 (3) (2015) 628–671.

[19] T.G. Selvaraj, B. Ramasamy, S.J. Jeyaraj, E.S. Suviseshamuthu, EEG database of seizure disorders for experts and application developers, Clin. EEG Neurosci. 45 (4) (2014) 304–309.

[20] K.R. Krishnan, S. Radhakrishnan, Hybrid approach to classification of focal and diffused liver disorders using ultrasound images with wavelets and texture features, IET Image Process. 11 (7) (2017) 530–538.

[21] K.P. Soman, K.I. Ramachandran, N.G. Resmi, Insight into Wavelets: From Theory to Practice., PHI Learning, New Delhi, 2010.

[22] R.M. Rao, A.S. Bopardikar, Wavelet Transforms: Introduction to Theory and Applications, Pearson Education Asia, Delhi, India, 1999.

[23] T. Ojala, M. Pietikainen, T. Maenpaa, Multiresolution gray-scale and rotation invariant texture classification with local binary patterns, IEEE Trans. Pattern Anal. Machine Intell. 24 (7) (2002) 971–987.

[24] S. Murala, R.P. Maheshwari, R. Balasubramanian, Local tetra patterns: a new feature descriptor for content-based image retrieval, IEEE Trans. Image Process. 21 (5) (2012) 2874–2886.

[25] Z. Guo, L. Zhang, D. Zhang, A completed modeling of local binary pattern operator for texture classification, IEEE Trans. Image Process. 19 (6) (2010) 1657–1663.

[26] R.M. Haralick, K. Shanmugam, I.H. Dinstein, Textural features for image classification, IEEE Trans. Syst. Man Cybern. (1973) 610–621.

[27] M. Abdel-Nasser, J. Melendez, A. Moreno, O.A. Omer, D. Puig, Breast tumor classification in ultrasound images using texture analysis and super-resolution methods, Eng. Appl. Artif. Intell. 59 (2017) 84–92.

[28] N. Nicolaou, J. Georgiou, Detection of epileptic electroencephalogram based on permutation entropy and support vector machines, Expert Syst. Appl. 39 (1) (2012) 202–209.

[29] J. Kennedy, R. Eberhart, Particle swarm optimization, Proceedings of ICNN95 – International Conference on Neural Networks, IEEE, Perth, WA, Australia, 1995.

[30] B. Xue, M. Zhang, W.N. Browne, Particle swarm optimization for feature selection in classification: a multi-objective approach, IEEE Trans. Cybern. 43 (6) (2013) 1656–1671.

[31] L. Boubchir, S. Al-Maadeed, A. Bouridane, Haralick feature extraction from time-frequency images for epileptic seizure detection and classification of EEG data, 2014 26th International Conference on Microelectronics (ICM), 14–17, IEEE, Doha, Qatar, 2014, pp. 32–35.

Index

Note: Page numbers followed by "*f*" and "*t*" refer to figures and tables, respectively.

Printed in the United States
By Bookmasters